How to Make It on the Land

How to Make It on the Land

BY RAY COHAN

GALAHAD BOOKS

NEW YORK

Reprinted by arrangement with
Prentice-Hall, Inc.

How to Make It on the Land
By Ray Cohan

Printed in the United States of America
Library of Congress Catalog Card No.: 73-90509
ISBN: 0-88365-131-9

To my wife, Hildegard;
In the tradition of Ceres,
as enduring and nurturant
as the land itself

CREDITS The illustrations and some of the material in Chapter 9, "Beekeeping," was furnished by the Agricultural Extension, Cornell University, Ithaca, N.Y.

The U.S. Government Printing Office in Washington furnished the material for Chapter 10, "The Family Forest," while the University of Alaska, Fairbanks, provided the information in Chapter 17, "Preserving Fish and Shellfish."

Original artwork was furnished by Martin Davis, Jr.

Material for driving a well in Chapter 5 was taken from the article "You Can Drive Your Own Well," adapted by permission; © 1970, Popular Mechanics, 224 West 57th Street, New York, New York 10019.

ACKNOWLEDGMENTS In the many months it took to gather and arrange the material for this book, I drew upon a lifetime devoted to an interest in the land. Translating that interest into a knowledge for others to follow and emulate, however, proved to be a humbling experience. For when the time came to set down that knowledge on paper, I realized how much I had forgotten and how I must now depend upon the experience of others.

And so I must express my indebtedness here in some measure, for either useful data, research material or other cooperation, to virtually every federal and state agricultural agency and Cooperative Extension Service of every state university in the United States and every important ministry, department and bureau in Canada. In addition to these federal and state research facilities, I owe much to the cooperation of private business corporations in both countries.

For want of space and with due apologies to the many I am unable to list, I mention:

The Agricultural Extension Services of the University of Arkansas; University of Alaska; Colorado State University; University of Connecticut; University of Missouri; New York State College of Agriculture of Cornell University; North Carolina State University and the North Dakota State University.

Also, the United States Department of Agriculture; United States Forest Service; and the departments of agriculture of the states of Delaware, Colorado, Florida, Iowa, Kansas, Maine, Montana, Nebraska, Pennsylvania, South Carolina, South Dakota, Commonwealth of Virginia and the Department of Agriculture, State of Wyoming.

In Canada I must thank The Ministry of Colonization; the Department of Lands, Forests and Water Resources, as well as the Bureau of Indian Affairs for the territories of British Columbia and the Yukon.

Contents

Chapter

How to Make It on the Land

I

The First Word

For the last five years I have listened to the voices of young and elderly America. For the last two I have traveled around the country to university campuses here and in Canada where I have listened to and rapped with students and faculty.

Out of the welter of rhetoric emerged one theme that seemed to bridge the generation gap apparent in both countries and which I for one was able to relate to intensely, because it brought my attention sharply back to a subject in which I have had a lifetime interest and much practical experience.

The theme was *land*; how and where to acquire it, and how to make it on the land one acquires.

This is a subject I know much about, an area where I can, without idle boasting, take a firm stance and meet the challenge of all comers. From the questions asked and answered during sessions on campuses from Columbia to Berkeley came my reason for writing this book.

My travel convinced me that the feeling, the hunger for LAND, is virtually universal, rather than endemic to a generation, class or economic level. It afflicts the young, the not-so-young, the elderly, the poor, the middle class and the very rich.

Anyone who fits into the last category is able to shop to his or her heart's content, smug in the knowledge that money is no object. This affluent breed, with a penchant to become gentlemen farmers with the latest of modern conveniences in house and barn, could read this book with profit—though they can well afford to pay for the mistakes it might help them avoid.

The primary target of this book is the individual, couple or group with little or no financial resources or experience in farming, little more, in fact, than a genuine yearning for the soil. This group, especially many of the younger ones who had already tried and failed to "make it" on the land, some as loners or with helpmates, others in one or another of the farm communes, provided much of the drive that led to this writing. Their abysmal ignorance of farming, their lack of orientation in preparation for the simplistic environment of the rural or wilderness scene, never ceases to amaze and shock when I consider their obvious intellect, healthy vitality, and above all, their spirited motivation.

It is aimed at those who, because of economic

necessity or principle, cannot or do not want to depend on any segment of the Establishment for their daily bread. Our goal is to show this group how and where to acquire cheap land and how to live on that land—successfully.

More, this book has been designed to show how crops and livestock can be raised successfully *without* the use of chemical fertilizers or insecticides and preserved *without* refrigeration. The raising of cash crops is also covered in depth. Included are detailed instructions and illustrations on preparing the soil, planting crops for livestock, personal use and winter storage. Other instructions will show you what livestock to raise. Nor does this book ignore fish and wild game management as a way to supplement both diet and cash income.

Other instructions deal with the construction of comfortable, low-cost do-it-yourself housing and other farm structures. There are detailed plans for building springhouses, smokehouses and root cellars—all absolute necessities where electricity and/or refrigeration is not used.

All structures described utilize materials produced by the land; they range from a simple log cabin to attractive rural homes made from wood and wood products, plans of which have been furnished by architectural experts.

More than a century ago, poet-abolitionist Herman Melville, commenting on the reelection of Abraham Lincoln to the Presidency in 1864, wrote: "The soil out of which such men as he are made is good to be born on, good to live on, good to die for and be buried in." Had land prices been as inflationary then as they are today, he might have added: *For those who can afford it.* Had he lived in the 1970s—with their environmental problems—it is unlikely that he would have written them at all.

The mythical land described in the opening chapter, "A Fable for Tomorrow," of Rachel Carson's book *Silent Spring* was once rich in wilderness and farm life. For centuries the trilling of birds in the forest, the droning of bees in the orchards and a cacophony of sounds from the barnyard had made it noisy in spring. Civilization and industrial progress finally arrived and smothered it in blight. Spring came and the countryside lay dying, without movement or sound. It was a silent spring.

Today's *real* countryside lies somewhere between the two extremes. Carson's countryside is still a threat for tomorrow, a warning of things to come if a national effort is not made to reverse the deadly, creeping paralysis that is strangling our heritage.

But, however bad the situation seems to us today, there still remain in this nation, and in that of our neighbor to the north, millions of acres of fertile soil and an abundance of clean air and sparkling water, begging for the hardy soul who is willing to cooperate with nature to make them bloom.

No state in our own nation matches the generosity of our neighbor to the north. Free land plus adequate financing are available at this writing in at least one province of Canada: Nova Scotia. Details on application by U.S. citizens for the free land are fully explored farther along. There is much cheap land in Canada. It ranges from uncleared to fully cleared land, from abandoned farms to working, fully-stocked ones. Priced far below comparable land south of the border, the amount of capital needed depends upon individual motivation and ambitions. A good rule of thumb to follow is this: The cost of land diminishes in direct proportion to the miles that separate it from major, populated areas, with the number of acres wanted a minor second factor.

Turning to the United States, where prices are invariably higher yet much good land is still available for those with limited capital, the key to buying LAND is still pretty much what it was thirty years ago. How far from urban centers is the buyer willing to settle? How few modern conveniences is he willing to settle for? What are his needs in acreage and what does he intend to do with them?

Alaska (where prospects for homesteaders remain dim) has a land-leasing arrangement which allows any U.S. citizen of nineteen years or over, five acres of "open-to-entry-land" in return for a filing fee of $10 and a rental cost of $40 per acre annually. However, Alaska *does* offer much cheap, unimproved land for sale.

In the United States various "dairy belts" offer the greatest chance for operating farms, abandoned farms and land for the buyer with limited resources. The reason that so many farms and so much land is available in these regions is due to a process of change from can to bulk pickup of milk that began about ten years ago and has since made "dairying" less profitable than it once was. Some sixty percent of farmers in the dairy belts have sold their herds. These belts are concentrated in the following regions, which once produced the largest quantity of can milk and still produce the greatest quantity of bulk milk: East North Central states, West North Central states, North Atlantic states, South Atlantic states, South Central states and Western states.

Wisconsin leads the nation in milk production, producing twice as much as Minnesota, which ranks second. New York is third, followed by Iowa, Illinois, California, Michigan, Pennsylvania, Texas, Missouri, Kansas, Nebraska, South Dakota, North Dakota, Oklahoma, Tennessee and Kentucky.

Include New England, the states of Washington, Oregon, Montana and Wyoming, add Canada's temperate Northwest territories and the temperate zones of her nine provinces stretching from the U.S. border north to the 55th parallel, and there is land enough for anyone at a price to fit any pocket.

But let us dispose of the land problem for a moment and return to my primary reason for writing this book.

I've made it on the land myself. By sharing my experiences with you, I can show you how to make it if you want to.

Let me say again that my goal is not to show anyone how to make it *big. Big* equals a successful commercial farm operation requiring substantial funding. I leave that task to the well-heeled businessman or syndicate who can afford to lose a little while learning the secrets of commercial farming from well-paid agricultural experts.

Instead, the goal as defined in this first chapter is living off the land. Basically this means acquisition of a quantity of proven or unproven land or an abandoned farm without livestock, tools or equipment and, usually, devoid of modern conveniences.

The most important factor here is not the number of acres, but the diversity offered. Plentiful water is a first essential. For general farming—a term you will find repeated often throughout the following pages—2 to 4 acres of clear, fertile land is more than enough to proceed with. Next in importance is a good-sized stand of mature timber (for exchange at the local sawmill for sawn, finished lumber for construction and/or repair of the house and outbuildings). Third in importance comes two hundred to one thousand hard-rock sugar maple, black or silver sugar maple trees for the first cash crop of the season, and enough second-growth timber for firewood and cover for wild game.

In hilly and/or mountain country other forms of natural building materials will be found—windrows of stone in fence rows, slate and shale. Rocky soil is a curse to the commercial farmer but a blessing in disguise to the small general farmer whose goals are shelter, food and cash—enough to pay his taxes and purchase other necessities that cannot be grown in the soil.

2

Land—U.S.A.

How to Make It on the Land was not written for those who prefer fairy tales to fact: It was written for people who are interested in the soil and are intelligent enough to realize that the act of creation requires effort on the part of the creator. It was also written for the millions of individuals who yearn for a piece of land they can retreat to or retire upon; a piece of land they may farm part time; a few acres that will furnish a substantial part of their living without their expecting it to give them everything they need on a silver platter.

It was written for those of you who wish to turn your backs on the Establishment and for those of you who do not; for people who dig communal living and those who do not; for the rugged types who can and will do without the conveniences of our modern age, as well as those who seek the tranquillity of "a place in the country"—but with all the comforts of "home." It was written for the young, the middle-aged and the old. Above all, it is aimed at the amateur farmer. The professional commercial farmer is really a businessman whose business is with the land he is helping to destroy as surely as his brother, the urban industrialist, is helping to destroy our cities.

Farming is far more significant than a method of making a part-time or full-time living. It is a mode of life, one that is completely foreign to the cities and the large towns.

You who become tillers of the soil will quickly discover that farming requires a major readjustment to a more simplistic environment, virtual nondependence upon the corner drugstore and supermarket and a full realization of the fact that you have exchanged your former frenetic existence for one of rural tranquillity and independence. Furthermore, the neighbors are farther away and there are all too few of the *real* conveniences to which city dwellers are accustomed.

Before You Buy a Farm

The slogan "Before you invest, investigate" is good advice for prospective farm buyers. When you purchase a farm or land upon which you will develop a farm, you will probably be making a long-term commitment—for better or for worse—so take your time about selecting it. Obtain the advice of several disinterested individuals. Personally inspect the farm and the section of the country in which it is located.

As a prospective buyer, you should investigate

the present development of the area; its probable future; the character and predominating nationalities of the people; the availability and kind of schools and community activities; and the kind of crops and livestock that experience has shown are best adapted to the locality.

Price

Don't pay too much for the farm or land. An overcapitalized farm or property can never pay reasonable returns on your investment. When estimating the net return from salable products, in order to establish a fair price, do not allow yourself to be misled by exaggerated reports. Instead, consult the advice of the local County Agricultural Agent.

The Down Payment

It is most unwise to put all or most of your money into a down payment. Enough capital should be set aside to pay living expenses until you have adjusted to your change of environment and decided to what income-producing use you are going to put the land.

Farmstead Water

The quality and quantity of domestic water available for household purposes as well as for livestock should be determined. Digging wells in semidry areas is expensive and the water supply may be inadequate during periods of excessive drought.

Location and Land Values

The farmseeker should settle in an area which experience has shown to be suitable to the type of farming he wishes to do. There are usually good reasons why certain crops or livestock predominate in a particular locality, and the new settler who attempts to operate contrary to local experience is probably headed for disaster.

The value of farmland is dependent upon a number of physical, economic and social factors, among which are soil type, fertility, topography,

climate, distance to market, buildings, community development, location with respect to schools and churches, commodity prices and type of farming to which it is adapted. These factors vary widely from area to area and from farm to farm. The price of farmland also depends upon the number of farms for sale and the demand for them. It is impossible to make a general statement that land in a given state, county, or locality is worth so much per acre. Even an experienced appraiser can only judge the approximate value of a property after a careful personal investigation of the farm and its surroundings.

When land is offered for sale at what appears to be an exceptionally low price, investigate the reasons. The price at which farmland is offered is, in general, an indication of its quality. Cheap land may be in an undesirable location or relatively unproductive, or it may be potentially productive but in need of heavy expenditures for improvements, such as drainage, clearing and buildings, before it is in condition for successful operation.

Size of the Farm and Its Establishment

The size of farm necessary to provide a family home and a reasonable income will vary so widely under varied situations and in the different states and localities that no set rule can be applied. There are districts under semiarid agriculture where 640 acres of land, 200 of which are in feed crops with a small acreage in cash crops and the balance in pasture, are barely adequate for a small family. In other highly cultivated sections 40 acres may be ample to provide a reasonably good income for a family if good farming methods are used and the operation is *highly diversified*. Consult the successful farmers in the locality to determine how many acres can be handled per man with good results.

The farmseeker must consider the following items in figuring what it will cost to set up a farm enterprise: (1) cost of farm and amount of

down payment; (2) living expenses until first or second crop returns are received; (3) expenses necessary to make the land productive, such as cost of fertilizer; (4) amount of livestock needed; (5) necessary machinery and equipment required; and (6) cost of repairing existing buildings or constructing new ones. Naturally, these costs will vary widely with different localities and different types of farm enterprises, but usually several thousand dollars of capital is required to establish a farm that has a reasonable chance of success. Too many people start farming with too little *capital, experience, and knowledge.*

Reclamation Farms

There are a few public farmland units, averaging about 80 to 120 acres in size, on some of the Federal Reclamation projects, which may be taken by qualified entrymen under the terms of the United States Reclamation Act. Settlers who take full-sized farms are required to have at least $4,500, unencumbered, or the equivalent, and at least two years of farming experience. Under a recent act of Congress a preference right of entry for not less than ninety days will be extended to veterans in connection with all reclamation openings. When an opening is announced, application may be made in writing to the project superintendent, but personal appearance before a local board is necessary before an applicant is approved.

Detailed information concerning these lands and the requirements that must be met can be obtained from the Bureau of Reclamation, U.S. Department of the Interior, Washington, D.C.

State

Certain state agencies issue lists of farms for sale or rent in their respective states. Land offices in some states can furnish information concerning state-owned lands for sale or rent. State rural credit offices may have farms to sell or rent.

Other Lands

Years ago it was possible to secure tax delinquent, abandoned and foreclosed farms. At the present there is practically none of this type of land available.

SOURCES OF CREDIT

Commercial Banks

All national, state and savings banks located in farming sections offer a wide variety of agricultural credit. All banks are familiar with farming in the areas they serve; many have farm departments or trained employees to handle agricultural business. Loans from these institutions are made to purchase farmland; to buy equipment and livestock; for repair, alteration and construction of new buildings; for fertilizer, land improvement and conservation; to finance the production, harvesting and marketing of crops; for breeding, raising, fattening and marketing of livestock and poultry; and to provide farmers and their families with cash for practically all purposes.

White River, Arkansas

Vermont

Federal Land Bank Loans

Federal Land Bank loans are made upon the security of first mortgages covering farmlands and the improvements thereon to borrowers who are engaged, or shortly to become engaged, in farming operations, including livestock raising, or the principal part of whose income is derived from such operations. This type of loan is made to purchase land for agricultural uses; to buy equipment, fertilizers and livestock necessary for the proper and reasonable operation of the mortgaged farm; to provide buildings and for the improvement of farmland; to liquidate indebtedness incurred for agricultural purposes; to provide funds for general agricultural uses; to pay living expenses and taxes.

Secretary-treasurers of national farm loan associations in all localities handle applications for land bank loans. If the secretary-treasurer cannot be located, write directly to the Federal Land Bank of the district.

Farmers' Home Loans

The Farmers Home Administration serves farm operators with needed technical help on farming problems. Its credit service provides loans for livestock, production and subsistence, farm ownership, soil and water conservation and emergency purposes. County offices of the Farmers Home Administration are generally located in the county seat towns and will supply application blanks for loans.

Production Credit Associations

The Production Credit Associations give credit to farmers to finance the production, harvesting and marketing of crops and livestock. Local or district offices are generally located in the more important agricultural counties.

Loans to Veterans

Under the Servicemen's Readjustment Act of 1944, as amended, war veterans are eligible for partial guaranty of loans which may be made to them by any national, state or private bank or Federal Land Bank, building and loan association, insurance company, credit union or mortgage and loan company that is subject to examination and supervision by an agency of the United States or of any state. Private individuals may also qualify as lenders.

Where the proceeds of the guaranteed loan, obtained by the veteran, are to be used for the purchase of farmlands, or in connection with farming operations, the following specific provisions of the Act must also be satisfied:

1. The farming operation must be "bona fide."
2. The property purchased with the proceeds must be "useful in and reasonably necessary for efficiently conducting" such operations.
3. There must be a "reasonable likelihood" of success.
4. The purchase price for such property must not exceed its reasonable value as determined by appraisal.

Further and more complete information may be obtained from the Administrator of Veterans Affairs, Washington, D.C., Veterans County Advisory Committees or from any bank, agency or other institution qualified to make these guaranteed loans.

Before signing a contract, make sure that the seller can furnish legal, clear title to the property. When considering purchase of land at a tax sale or from a state agency having tax deliquent farms for sale, particular attention should be given to the legal procedure governing such transactions and to the possibility of securing good title. The former owner often has certain opportunities for redeeming his property. In some states it is difficult to secure clear title from tax deeds.

When in doubt, get competent legal advice or the services of a reputable abstractor.

Physical Geography

The study of a good, fairly large-scale map of an area, showing locations of mountains, rivers, valleys, plains, plateaus, towns, roads, railroads, county lines, forests, parks, reservations and other features, will prove helpful in understanding the section. A relief map or a topographic map, which shows differences in surface levels or contour, is always helpful, particularly in an area where mountains or hills are common features. Variations in altitude or contour often cause striking differences in climatic and other conditions between places that are not far apart. General Land Office maps of the separate states and topographic maps of surveyed areas, prepared by the U.S. Geological Survey, may be bought from the Superintendent of Documents, Government Printing Office, Washington, D.C., at low cost.

Climate

Whether an area is warm, temperate or cold, humid or arid, cloudy or bright, windy or otherwise, may be determined from the United States Weather Bureau reports. The *Summary of Climatological Data for the United States by Sections* or the *Annual Report on Climatological Data* for the state in which you are interested may be consulted at various libraries throughout the country. The Weather Bureau in each state issues monthly and annual reports of the records of its observations. These may be obtained by writing to the proper state meteorologist.

Choosing the Ideal Farm

Before you choose the location of the farm or land you propose to purchase, you should decide what crops or livestock you are interested in producing. Make up your mind whether you want to chance a full-time or a part-time operation. I strongly advise the latter, within the following framework.

Part-time farming does not chain you to the land. It allows you to produce a massive percentage of your food without making you a slave to kitchen garden, field or livestock. It does require the operator to have some other source of income. This could range from retirement pension and/or Social Security to the sale of handcrafted objects, free-lance income from writing or artistic sources, part-time employment as skilled or unskilled labor, or the use of your land to take in summer boarders—to just name a few alternatives. Any one or combination of the alternatives suggested represents the heart of the general or *diversified* farm.

It is not really possible to discuss the size of the farm you should buy because of the endless factors involved, varying from individual taste and capital on hand to the particular area of the country in which you choose to settle. However, the ideal farm should include a minimum of the following requirements:

1. It should be located on a river, lake or stream.

2. It should consist of at least 20 acres of ground, approximately one-half wooded, of which the latter should contain a minimum of two hundred hardwood maple trees 9 inches in diameter or over. The unwooded section should consist of about 2 acres of pastureland, one acre of good garden land, approximately 5 acres of

good to fair crop and meadowland and 2 acres of miscellaneous land to provide room for human and animal housing, bee colonies and other purposes.

3. It should be ideally situated on or adjacent to a road carrying some out-of-neighborhood traffic, so as to take advantage of a road-to-farm market with a roadside stand for the sale of handcrafted items or excess produce. If you choose not to locate close to traffic, you should arrange to sell these items to retail outlets in large urban areas or through mail-order advertising. Don't knock such extra sources of income as boat livery service, picnic and trailer or campsite grounds and taking in summer boarders and hunters during the game season. For other sources of income see the appropriate chapter.

Statistically Speaking

I'm not going to bore you with a lot of figures, but you should know how much housing space, feed and/or pasturage you will need for livestock, acreage for home garden and fruit production, costs of livestock and figures on the yield you may expect in return for your investment and labor.

For Example

One Dairy Cow: You will require approximately 200 square feet of floor space, including storage for hay and grain. The typical dairy cow will produce two thousand to four thousand quarts of milk each year and one 85-pound calf every 305 days. She will need one-quarter ton of grain, costing $60, 3 tons of hay worth $35 per ton, and 2 acres of pastureland; miscellaneous items of bedding, breeding and veterinary fees will cost an additional $25 per year. Your original investment in a dairy cow with calf or an animal in milk will range from $150 to $300, according to locality and breed. I personally recommend the Jersey cow because of the high butterfat content of her milk. If you are not interested in butter, cheese and other dairy products, choose a Holstein or Guernsey, which produces a great deal more milk but of a lesser grade.

Sheep: Minimum housing against cold and wet weather may consist of a simple open shed. Pasturage requirements for three adult sheep and their lambs come to one acre of good to fair ground. You may expect 8 pounds of prime wool from each sheep annually and each lamb will weigh 85 pounds before slaughtering and butchering: net yield of prime lamb—45 pounds. Feed 50 pounds of grain and one-quarter ton of hay to each sheep annually. Keep them behind a stout fence of woven wire for protection against dogs.

Pigs: One of two pigs annually will take care of the pork requirements of the average family. Pigs are usually purchased as weanlings for about $10 each. They will eat garbage, skimmed milk, and garden surplus, in addition to 400 pounds of grain per animal until they are slaughtered in the late fall.

Beef: A beef calf raised to a 700-pound yearling will require about 800 pounds of grain, a ton of hay and an acre of pasture.

Rabbits: A buck and three does will produce all the rabbit meat required by most families. Rabbits must have dry, sanitary quarters. Feed requirements are small but good-quality hay and greens are needed.

Goats: Two or three goats will produce enough milk to meet the demands of the average family. Goats require clean, dry quarters and feeding rations similar to sheep. The milk is subject to "off-flavor" unless it is carefully handled. A buck must be available for breeding.

Laying Hens: About 3 square feet of floor space are required per layer. Each hen will lay about sixteen dozen eggs in a year's time and consume 110 pounds of feed, costing about $5 per hundred pounds. One out of every four hens will die or be culled for nonproduction each year. Replacements can be purchased at a cost of about $2.25 each or hens can be raised from baby chicks. This latter procedure requires special housing and equipment and is not recommended for a family flock of ten to twenty hens.

Vegetable Garden: A half acre of garden can be handled without specialized tools or equipment, providing plowing is hired. This size garden will produce about all the fresh and storable vegetables most families can manage. These would have a retail value of about $125.

Small Fruit: Four grape vines, twelve blueberry bushes, 100 square feet of strawberry bed, 50 feet of raspberries and ten assorted fruit trees will provide the average family with all its fresh and storage fruit needs. These crops need specialized equipment for spraying and dusting.

Remember, the above figures are approximate and vary from state to state.

Before I leave you for the next chapter, let me say that it is impossible to give accurate figures on land costs in the United States per acre because of the enormous discrepancy that exists from one state to another and even from one county to another within a particular state. The best I can do is give you the cost of an average acre of excellent to fair to good land in the Northeast as against costs of uncleared land in the Far Northwest *without* buildings.

Excellent farmland in the Northeast ranges from $500 to $1,000 per acre in 200- to 400-acre lots and from $3,000 to $5,000 per acre in lots of only one to 3 acres. Good to fair land ranges in cost from $350 to $500 per acre in 200- to 400-acre lots to $1,200 to $2,400 in lots of one to 3 acres. In the Northwest, good uncleared forest land can be had from approximately $100 per acre and up, according to state and locality.

Unfortunately there simply is not room enough here to give a state-by-state breakdown of land sources beyond what I have already listed on previous pages, but if you write to the National Institute of Farm and Land Brokers, 155 East Superior Street, Chicago, Illinois 60611, telephone (312) 664-9700, you will receive a national listing, geographically and alphabetically arranged, of all accredited farm and land brokers in the United States and Canada. Happy hunting!

3

Land—Alaska

Because of the special conditions that exist in Alaska, our forty-ninth state, I am treating it as a separate entity and chapter.

Alaska gets its name from the native word *al-ay-ek-sa*, meaning "the great land" or "mainland." It is one-fifth the size of the continental United States, contains 586,400 square miles and is as large as Texas, California and Montana combined. The 1966 population was 272,000, an increase of 24.5 percent since 1960. About one-sixth are Indians, Aleuts and Eskimos.

Farms are scattered from Juneau in the southeast to Kotzebue in the northwest, but most are in the rail-belt area which includes the Tanana Valley, Matanuska Valley and Kenai Peninsula. This area produces nearly ninety percent of the total agricultural value. Most farm products are marketed in Anchorage, Fairbanks and surrounding military installations.

Dairying is the most important farm enterprise. In terms of agricultural commodity production, potatoes, eggs, grass hay, grain silage, beef and veal, reindeer, and wool follow in that order of importance. Beef and sheep are raised primarily on Kodiak and adjacent islands in the Aleutian chain, where mild weather allows a long grazing season. Reindeer are raised in the arctic tundra area on the Seward Peninsula.

Alaska is best known for its large, high-quality vegetables which flourish in the cool, moist summer months, along with the long hours of sunshine. In this climate smooth bromegrass and a mixture of oat-pea grow fast and yield well. These latter grasses are Alaska's best forage crops in terms of value and acreage.

Alaska is a large island with wide differences in climate. Southeastern Alaska or the "Panhandle," Kodiak Island, the Aleutian Islands and much of the Kenai Peninsula have a maritime climate with warmer winters and cooler summers than in the interior. The growing season is usually longer, but cloudiness, abundant rain and cool temperatures may delay crops from maturing.

Farther north, in the interior, summers are shorter but warmer and winters are colder. Frost-free periods in the Tanana and Matanuska valleys are about one hundred days. In spite of the shorter growing season, however, the long days of sunshine and warmer temperatures make crops mature faster. In summer Anchorage has up to eighteen hours of sunshine daily, and

Two miles south of Palmer, Alaska

Fairbanks has twenty-one hours. April, May and June are often the driest months. Rains usually come in late summer and fall, making haying difficult and slowing ripening of grain.

Land over 2,000 feet elevation is not recommended for farming since freezing can occur any month of the year. Narrow valleys and low hollows often have inadequate air drainage, increasing the risk of frosts. On northern slopes snow melts slower and soils warm up later. A high ridge or mountain to the east will block early morning sun and shorten hours of sunshine. In some areas strong winds blow fields bare of snow and cause winter damage to perennial grasses and legumes. The best locations for farms are gentle southern slopes and wide valley floors having adequate air drainage. Conditions vary even between adjacent homesteads. The settler should check land locations very carefully.

Summer is the best time to select land. In winter a blanket of snow covers the frozen ground, making it difficult to tell the swamps from the small flat valleys. Yet almost one-third of all homesteads filed in recent years were selected by persons seeing the land only under snow. People usually won't buy a house without first seeing it, yet they will select a homestead without a true appraisal, which often leads to failure.

If the soil can't support a good growth of native trees or grasses, it usually won't produce good crops. Flat ground covered with moss, low bushes and shrubs is generally muskeg with poor drainage, and the frost line is often near the surface. Flat areas with stands of small black spruce are usually poor farmlands. The "swamp spruce" are normally less than 20 feet high and 5 inches in diameter, and grow in dense stands. Permafrost is often a foot or so below the surface, with drainage poor.

Mature stands of birch and spruce are normally found on good soil with adequate drainage. Cottonwoods grow where there is adequate surface drainage; the topsoil—while often good—

may be quite shallow, with a subsoil of pure sand and gravel. Alders and willows generally grow on river bottoms where water is near the surface and drainage may be a problem.

The natural cover, however, is only a clue to underlying soil. Take time to check soil depth; dig down 4 or 5 feet. Soils less than 2 feet deep are questionable for farming. Layers of heavy clay at shallow depth may prevent adequate drainage, and shallow soil over gravel may dry out too quickly.

Permafrost is a condition where the soil remains frozen all year round. It is dependent upon the geographical location of the land, the direction of the slope, the type of vegetation covering the land, and the elevation. Usually permafrost does not exist where the annual yearly temperature is 32° F. or above. Permafrost creates problems for farmers because the soil is cold and has no internal drainage. Homeowners have serious problems in building, obtaining water and disposing of sewage on land underlain with permafrost.

In the Tanana Valley some of this land has been made into farmland by clearing the vegetation and letting the land lie bare for two to four years. The object of removing the vegetation is to remove the insulating cover over the soil. With the cover gone the frost recedes. In some places the permafrost is really a field of ice lying under the topsoil. When this ice below the surface melts it can create huge pits; if the ground settles over these pockets of melted ice the fields will be very rough.

In the Matanuska Valley most good farming land is privately owned. Land still available for homesteading is usually remote from roads, electricity and telephones. Dairying is the most important farm enterprise, with potatoes second. Good growers average 10 to 12 tons of U.S. No. 1 potatoes per acre. Efficient vegetable growers can get 12 to 15 tons of cabbage, 7 tons of lettuce and 7 to 9 tons of carrots to the acre. The growing season ranges from 100 to 120 days. During the season 7 to 8 inches of rain falls and because of low evaporation rates it is

usually adequate. Late summer rains hinder haying but May and June are usually dry.

The Lower Susitna Valley has similar soils and growing conditions. It consists of uplands and river terraces, interspersed with muskeg. The Upper Susitna Valley has short frost-free seasons, with only sixty- to seventy-five-day growing periods in some areas.

The Tanana Valley, near Fairbanks, is the second largest farm area. There are more potential agricultural lands available there than in the Matanuska Valley. Contouring is especially important because the high mica content in soils contributes to erosion. Land-clearing costs are somewhat less than in the Matanuska Valley. The growing season is about one hundred days, with long hours of sunshine and high temperatures.

In the Kenai Peninsula most farms are small and in the beginning stages of development. Much of the food produced is locally consumed. There are a few dairy and beef operations and other types of small farms. Much of the area south of the Kasilof River is covered by dense native grasses and vegetation, offering potential grazing for three to four months. Winter grazing is usually impossible due to damp cold weather, heavy snow and the low feeding value of winter vegetation. Some native grass is cut for winter feed. Climate is maritime with ample precipitation and cloudy weather. Temperatures during growing season average 44°–61° F.

In southeastern Alaska the chief products are fish and timber. The Forest Service estimates that an annual harvest of over a billion board feet of timber for plywood, pulpwood, lumber and by-products can be harvested perpetually from the 21 million-acre National Forest. Climate is maritime and precipitation heavy.

On Kodiak and the Aleutian Islands, stock raising (sheep and cattle) is the leading agricultural enterprise. Natural vegetation and grasses provide seasonal and even year-round grazing, although supplemental feeding is recommended. Open rangeland is leased from the federal government at modest rates. Expansion is limited not by rangeland, which is ample, but by lack of winter feed, transportation costs and poor marketing opportunities. Wool is shipped to the States. Most islands are treeless but some are unsuitable for livestock raising as they are too rugged and have no beaches or harbors. Climate is primarily maritime with heavy precipitation. Some beef is exported to Anchorage, and a few sheep and cattle are killed for local use. Slaughtering facilities are badly needed. The marketing structure on these islands is undeveloped. Transportation distances are long and the costs are high. These factors have caused some ranchers to fail and in general have slowed down the agricultural development on these islands. The only farm product exported in any quantity is wool from the Aleutian Islands.

Homesteading

Selecting land is the most important thing a homesteader does, for land and its location determine clearing costs and crops that can be grown. They also determine availability of schools, neighbors and community services. If the settler lives in the lower states, he should invest in a trip to Alaska to see firsthand the conditions he will encounter and the location he may want.

Free Land

The term "free land" is misleading. Under federal homestead laws, total costs for filing for 160 acres of land vary from $50 to $260 depending on whether the claim is initiated by settlement or application and the type of final proof filed. The homesteader must also pay the cost of publishing notice of final proof. Expenses of clearing land, building a house, planting crops and other costs must also be taken into consideration.

Basically there are three methods of acquiring land: filing under the various federal public land laws, buying private land, or homesteading (buying) state land.

Homesteading laws were designed for agricultural development. If a person just wants a place

Colville River delta, Alaska

to live, he should consider a homesite settlement claim on federal land or buying a small tract from the state government or a private person.

It is not possible to file for land by "remote control" in Alaska. Public land laws generally require occupancy or personal examination of lands before filing.

There are land locators who will examine land status, aid in finding land and fill out applications for a fee. They receive the same information and maps available to anyone interested in filing an application. A person using such a service should make certain he is dealing with a reputable organization.

The Geological Survey, U.S. Department of Interior, sells maps in two scale sizes showing roads, lakes, streams, elevations and survey boundaries. The Soil Conservation Service has maps of major agricultural areas showing soil characteristics and types, streams, roads and section lines.

If a person files for unsurveyed land, he must stake each corner of his tract, laying it out as nearly rectangular as possible. Homesteading narrow strips of land along waterways and lakes is not permitted. Natural boundaries such as lakes or rivers may be used, but other lines should run straight and true north and south, east and west where possible.

An agent may represent you at a land auction, but he may represent only one person.

The chief advantage in homesteading is avoiding an initial large cash outlay for land. The larger cost of development is spread over a number of years. Until development is made, however, almost no income is earned from farming.

The person filing must be a U.S. citizen or have formally declared his intention to become one, be at least twenty-one years old or the head of a household. Unmarried women may homestead. Under certain circumstances a married woman may also, but a husband and wife may file only one homestead between them. Up to 160 acres may be homesteaded.

Homesteaded lands are subject to local property taxes before a patent is issued, but are usually not taxed. Personal property and improvements are taxed, however. There are a state income tax, motor fuel tax, sales tax and other local taxes commonly found elsewhere.

Newcomers may face such hardships as the remoteness of the homestead from towns or other persons; lack of schools, hospitals, electricity, telephone and other community facilities; and initial lack of plumbing and water. A wise settler should make certain his wife fully understands the conditions she will face and the hard physical work necessary both indoors and outdoors due to lack of modern conveniences. Her attitude and ability may well determine the family's success.

Cabin Sites on Hunting and Fishing Lands

The Land Office of the Federal Bureau of Land Management has topographic maps for counter use. Finding the particular cabin site will help in looking up records to see if the land is privately owned, withdrawn by the government, selected by the state or unclaimed. If unclaimed, check with the nearest Land Office. If the site is on state land, check with the Division of Lands office in Anchorage, Juneau or Fairbanks.

A homesteader without patent to his land would have trouble borrowing money though it is not prohibited by the homestead law, whereas a settler with purchased land may be limited in ability to borrow until he has buildings and land ready for production. Most lending agencies require farming experience in Alaska for eligibility. Private banks are often reluctant to make agricultural loans: These are short term, usually for three years or less.

The Alaska Agricultural Loan Board (a State agency administered by the Alaska Division of Agriculture) and the Alaska Rural Rehabilitation Corporation both have authority to make long-term loans; due to limited funds, however, these loans are usually for ten years or less. The latter corporation, headquartered in Palmer, has made several long-term real estate loans through the Federal Land Bank program, but these are based primarily on land value and do not apply to a homesteader.

The Farmers Home Administration is the largest source of farm credit in Alaska. It makes several types of loans, ranging from emergency and operation loans for short periods to ownership loans up to forty years. FHA farm loans are restricted to persons who earn their major income from farming and who are unable to obtain needed credit elsewhere. The state office is in the Arctic Bowl Building in Fairbanks.

The Agricultural Conservation Program provides cost sharing for certain soil and water conservation practices and also supplies grain storage and dryer loans. Small pieces of land for rural housing are sometimes available along maintained roads, usually from private owners. Good farmland is usually occupied even before major roads are completed. Most homesteaders have the choice of taking poor land along an existing road, or good land and building their own road.

Most settlers have to build access roads to their site, and cost varies tremendously. To clear a road right-of-way with a hired bulldozer, in an average wooded area, may cost $300 to $1,000 per mile. If you put gravel on your private road this will be an additional expense. If ditches are crossed in tying in a private road to a public one, approved culverts or bridges must be used. Remember also that on a private road the settler does his own maintenance and snow removal.

Land clearing is usually the largest single expense. Clearing by hand is cheap but slow and yields a poor return for labor. Bulldozing is the usual method, with brush piled into boundary fences. Clearing land usually costs $50 to $250 an acre, depending on size of trees, density of brush and timber, size of equipment and skill of operator. This does not include picking roots or preparing the land for cultivation. Careless clearing results in loss of large amounts of topsoil; in Alaska the formation of new topsoil is extremely slow.

Drinking water is a necessity. In a few places homesteads have been abandoned due to an inadequate water supply or the high cost of drilling a well. Some settlers haul drinking water until a well is drilled. Cost for drilling may be from $7 to $13 or more a foot. Some settlers hand-dig wells, but this can be dangerous and time consuming. If an area has a history of dry wells, a settler will be wise to drill a well first to be sure he has adequate water.

REA (Rural Electrification Agency) cooperatives service major areas of the Kenai Peninsula, Matanuska Valley and Tanana Valley. In more remote areas electricity is usually not available unless small private power plants are used. There is rural telephone service in the Matanuska Valley and certain areas around Homer and Fairbanks. The Army Signal Corps provides some service along major highways and the Alaska Railroad.

Larger cities all have modern schools. The state usually provides grade school teachers where fifteen or more students are not served by an existing facility. The state may build a schoolhouse or it may provide only a teacher, with the community supplying a building. School bus service is usually made available where necessary. The state also has an excellent correspondence course for children too isolated to attend school. This course can be taught by anyone of average education.

The Matanuska Valley Farmers Cooperating Association in Palmer has a complete stock of farm supplies and equipment. Stores in Fairbanks, Homer and Anchorage also carry most needed farm equipment. Popular makes of farm machinery and equipment are sold at regular list prices plus cost of transportation. Brand name feed and fertilizer are also available. Some settlers locate near areas where they can rent heavy equipment such as bulldozers and special plows to prepare newly cleared fields.

Vegetables in Alaska

Approved varieties suitable for Alaska's climate include potatoes, cabbage, carrots, beets, celery, lettuce, broccoli, brussel sprouts, peas, cauliflower, rhubarb, turnips, radishes, strawberries and raspberries. Tomatoes and cucumbers grow well in greenhouses. Potatoes are the second major source of farm income in Alaska. The military buys most of these potatoes and because growers compete for the market, prices are often below those of imported potatoes. Agricultural Experiment Stations constantly test and develop varieties of vegetables and crops suited to Alaska. The Cooperative Extension Service has circulars listing the recommended varieties.

Recommended Field Crops

Smooth bromegrass is the principal forage crop used for hay, silage and pasture. Perennial legumes do not survive well. Oats can produce 2 tons of hay and 6 to 7 tons of silage per acre. Cool, wet fall weather often makes it necessary to dry hay and grain artificially.

The settler should always choose varieties tested and found adaptable to Alaska. A good initial crop for homesteaders is oats, or peas and oats cut for hay or silage. Grains such as barley and oats are good if harvesting equipment is available. Potatoes and vegetables are recommended for first-year crops as a family garden.

Freshly cleared Alaskan soils are less productive than those in most states, and to get good yields fertilizer must be used. Soil can be tested through the Cooperative Extension Service to determine how much fertilizer is needed. Alaskan soils normally require 200 to 400 pounds of mixed fertilizer per acre for field crops and 750 to 1,000 pounds per acre for vegetables annually.

Control measures must be used throughout Alaska against root maggot, which attacks vegetables such as cabbage, turnips and radishes. Cutworms are also a nuisance. The most serious economic plant disease is ring rot in potatoes. Mosquitoes are bothersome at certain seasons but do not carry disease locally. Dairy herds are tested annually for tuberculosis.

Sheep and beef production is centered on Kodiak and adjacent islands, and in coastal islands of the Aleutians. Here mild weather permits a long grazing season. Many herds graze year round, although supplemental feeding is recommended. Most of the land is leased from the government. The number of sheep and cattle has doubled in the past five to ten years. Presently there are about 14,000 sheep, mainly on the Aleutian Islands, and 2,500 cattle, mostly on Kodiak and Chirikof islands.

Living Off the Land

A good hunter and fisherman can obtain moose, rabbits, bear, caribou, ptarmigan and grouse, grayling, and salmon, depending on his area. Game, however, is not always readily available and may take considerable time and energy to secure. Until the settler has been in Alaska a year, he must purchase nonresident hunting and fishing licenses. Newcomers to Alaska should not plan to "live off the land."

There are about 42,000 reindeer located in western Alaska. Only natives may own reindeer herds. There are fifteen to twenty privately owned native herds, but the main source of commercial reindeer meat is from a government-managed herd on Nunivak Island.

State law provides that "timber land shall not be disposed of by sale or lease." The timber itself, however, may be sold in such a way as to meet long- and short-term needs of forest industries. A sale up to a maximum of 500,000 board feet may be negotiated, and an individual is limited to one negotiated sale per year. Larger sales must be competitive.

An individual desiring to cut timber for construction of a house, barn, fence or other outbuildings for his own use must obtain a Personal Use Permit. No charge is made for this timber on state-owned land.

In areas where building materials are scarce, a homesteader must often build with logs. Less cash may be required but log construction takes more time and hard work, and good logs may be hard to find. A good weathertight fit takes considerable skill and several men are usually needed to raise a cabin. Lumber is faster and easier to work with and the average person will probably build a tighter, more weatherproof house. Most homesteaders build their own homes since labor represents forty to sixty percent of the cost of building an average house. A set of plans can prevent costly mistakes and usually results in a better building.

Yukon River, Alaska

It is more expensive to live in Alaska than anywhere else in the U.S. Living costs range up to fifty percent higher, due primarily to transportation costs, high wages, construction costs, high processing and distribution costs. Some of the adverse effects of high living costs are offset by higher wages. The economic pressures, however, are still great for many families.

In December 1966 a basket of groceries containing forty food items and retailing for $17.25 in Seattle cost consumers in Juneau $21.45, in Anchorage $22.09, in Fairbanks $24.20, in Nome $28.87 and in Ketchikan $21.11.

A family considering a move to Alaska should have a cash reserve of at least $800 after their arrival. This money will be needed to pay for the rent, food, clothing, auto expenses and other necessities of life that are required in the first month. Many families underestimate these resettling expenses.

Employment opportunities for outsiders are scarce, seasonal and specialized. Shortages do exist in a few occupations at times. Information about job opportunities can be had from State Employment Security agencies in major cities in Alaska.

Farm Employment

Many farmers make the major share of their income at off-farm jobs. Such jobs, however, are usually for summer months when the farmer should be home tending crops and developing land. It usually takes five to ten years before a homestead can support a farm family, and many homesteaders fail because of lack of money and opportunity for off-farm work. A farmer must usually be highly skilled or educated to compete in the employment market. A would-be farmer should consider availability of off-farm employment in the area he is choosing.

Living conditions and customs in Alaskan cities and towns are similar to those of equal size in other states. There are modern stores,

churches, schools, medical facilities and public utilities. Rural areas are much more primitive. Information on specific cities can be obtained by writing to the Chambers of Commerce in those cities.

The Alaska Highway

The Alaska Highway is kept open year round. From the Canadian border to Fairbanks the highway is blacktopped; only limited portions are paved on the 1,221 miles from Dawson Creek, British Columbia, to the Alaskan border. Each year a few more miles are paved, but as yet most of the Alaska Highway in Canada is gravel. The gravel portions are well-graded and banked, although there may be local bumpy areas due to annual freezing and thawing. Accommodations, gas, oil and repair service are available at many places along the highway.

There are about 4,000 miles of highway system, with 1,000 miles paved or blacktopped and the rest dirt or gravel. About 2,000 miles of this system are maintained summer and winter. A primary road system leads from Seward and Homer in the south to the northernmost point of Circle on the Yukon River.

Public Campsites

A publication, *Camping Under the Midnight Sun*, is printed each spring and gives location and types of state-operated facilities available to the traveling public. This booklet can be obtained from the Alaska Division of Lands in Anchorage. Not listed in the booklet are the many campgrounds in the Bureau of Land Management system.

Transportation

A new state ferry system operates between Prince Rupert, British Columbia, and Haines and Skagway in Alaska. Skagway connects with Whitehorse on the Alaska Highway via the Whitepass and Yukon Railway. From Haines, the traveler can drive the 154-mile Haines Highway cutoff to Haines Junction on the Alaska Highway, 96 miles east of Whitehorse. The three ferries can each carry five hundred passengers and over one hundred cars; they have a dining room, snack bar, cocktail lounge and staterooms with upper and lower berths. The complete trip takes about one and a half days.

There is a daily jet service to and from the lower states, and a well-developed air transportation system operates within Alaska itself. Over-the-Pole-Flight has made Alaska a major hopping-off point to Europe, and many flights stop in Alaska en route to the Orient.

The Alaska Railroad, owned and operated by the federal government, provides service from Seward north to Fairbanks. It is about 500 miles long, with a branch into the Matanuska Valley, and makes regular runs between Fairbanks and Anchorage.

Water Rights

Anyone who uses water and wishes to continue using it, whether to supply a small household or a city or industrial plant with millions of gallons each day, must file for water rights.

The Division of Lands issues these rights upon application. The application, a declaration of appropriation, is a formal statement from the water user giving the location, quantity and history of the water use.

An existing water right, or a "grandfather right," is a water right for the use of water that was effected before, or was under construction, or was used within five years prior to July 1, 1966, the effective date of the Water Use Act.

For Further Information

Agricultural Stabilization and Conservation Service
Room 413, 516 Second Avenue
Fairbanks, Alaska 99701

Bureau of Land Management
Cordova Building
555 Cordova Street
Anchorage, Alaska 99501

Farmers Home Administration
Arctic Bowl Building
954 Cowles
Fairbanks, Alaska 99701

Soil Conservation Service
Severns Building
P.O. Box F
Palmer, Alaska 99645

State Division of Agriculture
Box 800
Palmer, Alaska 99645

State Division of Lands
344 Sixth Avenue
Anchorage, Alaska 99501

University of Alaska Agricultural Experiment Station
Box AE
Palmer, Alaska 99645

University of Alaska Cooperative Extension Service
University of Alaska
College, Alaska 99701

U.S. Forest Service
Regional Office, P.O. Box 1631
Juneau, Alaska 99801

4

Land—Canada

It may come as something of a shock to people who have never been there but, basically, except for the Arctic regions, the high mountain country in the far west, and the Yukon, Canada is *not* a land of "snow and ice." In fact, year-round temperatures average only slightly below those of our own northern states.

Without attempting to sell short the United States, Canada, for the young, the middle-aged and the elderly, can be the more hospitable nation with a better and more relaxed political and social climate. The world's largest nation, with the exception of the Soviet Union, Canada has a special regard for Americans and readily welcomes them as citizens.

Once your application has been processed you are immediately eligible for socialized medicine and old-age pension (if you have reached retirement age). You are also immediately entitled to any and all other advantages enjoyed by native Canadians because you have automatically become a citizen of that country. Other benefits include, in some provinces and territories, the option of buying or leasing some of the world's most spectacular land—at very low cost.

Up to a few months ago, Canada was offering *free* land plus financial assistance to homesteaders. Today that Crown land is depleted. No province or territory any longer has free land to dispose of—but low-cost land prices plus generous long-term leasing, especially in its Northwest (British Columbia) and Yukon territories, still make our neighbor to the north a prime land prospect.

As a good example of what I mean, let's take the Territorial Lands Regulation Act, enacted in 1961, which governs the disposal of territorial lands in British Columbia and the Yukon Territory. For the purchase or lease of such land, preliminary fees total less than $50 U.S. Leased land, giving the lessee surface rights to the property for thirty years with a renewal option for thirty years more, requires an initial investment of exactly $10 Canadian.

Application for permanent residence in Canada can be made simply by writing the Department of Manpower and Immigration, Central Processing Office, P.O. Box 75, Ottawa 2, Ontario, Canada.

Provincial Crown lands can also be purchased or leased at low costs. A further noteworthy provision for prospective purchasers of provincial and territorial lands is a regulation pertaining to the leasing of campsites for the settler who wishes to take his time regarding homesteading, purchasing or leasing farmland sites of adequate acreage. Consisting of an acre or less and costing $45 annually, including the $25 processing fee, leasing requires the construction of a cabin with a minimum floor area of not less than 168 square feet. Since you are free to cut materials from Crown lands, construction costs are held to a bare minimum. For application and further information, write the Minister of Natural Resources in the province in which you are interested in settling.

Geography and Climate

Stretching 5,780 miles (9,284 kilometers) from east to west, Canada covers an area of 3.8 million square miles (9.7 million square kilometers) in which lives a population approaching the 22 million mark. Canada, comprising ten provinces and two territories, exceeds the area of the United States by more than 200,000 square miles (512,000 square kilometers). To cross it by train takes four days and four nights; to fly from Montreal to Vancouver, on Canada's west coast, takes as long as to fly from Montreal to Paris.

Nearly half of Canada is covered by both forests and the Canadian Shield, a 1.8-million square-mile (4.6 million square kilometers) area of ancient rock, mainly in the north which contains immense mineral wealth. Mostly a region of hills, lakes and muskeg or swamp, the Shield is sparsely populated. By comparison, almost 70 percent of Canadians live in urban areas concentrated within 100 miles of the United States border. The fresh water which forms more than 7.6 percent of the total area of Canada represents about a quarter of the world supply.

The Country Regions

Canada is divided into five basic regions: the Atlantic provinces, Quebec and Ontario, the Prairies, the Pacific Coast, and the North.

Atlantic: The four Atlantic provinces of Nova Scotia, New Brunswick, Prince Edward Island and Newfoundland all have extensive coastlines and the inhabitants owe much of their livelihood to the sea. The first recorded landing in Canada was made in 1497 in Newfoundland by John Cabot, an Italian-born explorer in the service of England. The Vikings, sailing from Iceland, are believed to have reached the coast of the Atlantic provinces several centuries earlier. In Labrador, which is part of the province of Newfoundland though it lies on the Canadian mainland, the world's largest underground power station is being built to harness Churchill Falls. When completed, this hydroelectric development will generate 5.25 million kilowatts. Fishing, farming—including fruit growing—coal mining, pulp and paper, and manufacturing are the predominant industries of the four provinces.

Quebec and Ontario: The central provinces of Ontario and Quebec contain more than half of Canada's population. Their industrial growth has been favored by the St. Lawrence River and the Great Lakes, which constitute one of the world's great waterways and carry deep-sea shipping more than 2,280 miles (3,648 kilometers) from the Atlantic Ocean into the heart of the continent.

Quebec is one of the most important industrial provinces and mining is its most important primary industry. Hydroelectric power ranks second as a primary resource. Quebec's forests provide thirty-eight percent of Canada's pulp and forty-three percent of its paper production.

Ontario is the foremost industrial province of Canada but it also produces about a quarter of Canada's net income from farming. Mining is of great importance: Most of Canada's nickel and platinum, and much of its gold, copper, cobalt and salt, are mined in Ontario. Niagara Falls, one of the largest waterfalls in the world, is the chief source of hydroelectric power in the province.

Prairies: The three prairie provinces of Alberta, Saskatchewan and Manitoba are covered with rich black earth upon which much of the world's grain is grown; however, each year industry assumes greater importance in the economies of the three provinces. Saskatchewan has a large

potash industry and Alberta's oil fields produce sixty-three percent of Canada's oil and eighty-five percent of its natural gas. Manitoba is an important producer of copper, nickel and zinc.
Pacific Coast: More than seventy-three percent of the westerly province of British Columbia is forested, and it is the country's largest producer of lumber. Most of the province is covered by mountain ranges which run from south to north. In the south of the province are a number of interior valleys which contain some of the finest fruit-growing districts in North America. British Columbia is rich in hydroelectric power, using only one-tenth of its potential of 30 million horsepower. In metal production, the province ranks fourth among the other provinces with important deposits of zinc, lead, copper, silver and gold. Another important resource is fish, particularly the Pacific salmon.
North: Northern Canada is the last remaining frontier. Through its vast mineral resources it holds the key to Canada's future. Northern Canada is made up of the Northwest Territories, whose surface of 1.3 million square miles (3.3 million square kilometers) covers more than a third of Canada including the Arctic archipelago, and the Yukon Territory.

Year-round oil exploration is under way in the Canadian Arctic and has already resulted in significant finds. Studies are being made to find economic methods of transporting the oil south. An estimated one-third of Canada's fresh-water supply is in the north, much of it in rivers flowing into the Arctic Ocean. The combined population of the Northwest Territories and the Yukon Territory is 46,000.

Climate

Canada's climate is characterized by marked changes of weather with the changes of season. The seasons run as follows: spring–mid-March to mid-May; summer–mid-May to mid-September; autumn–mid-September to mid-November; and winter–mid-November to mid-March.

Summer in Canada is usually very warm and frequently humid. Swimming, sailing and all types of outdoor activities are enjoyed by Canadians. In all but British Columbia, winter is cold and snowy but often marked by long hours of sunshine. Coastal British Columbia is tempered by warm Pacific air; winter temperatures are similar to those in the United Kingdom.

Newcomers to Canada soon learn to cope with winter, which for many–particularly those who enjoy winter sports–is one of the most enjoyable seasons. Homes, offices and factories are centrally heated in winter, as is all public transport. Most city streets and country roads are open to traffic within a few hours after a major snowfall. A new development in some urban areas is the construction of vast, heated indoor shopping centers where customers may shop in comfort during the coldest weather.

Time Zones

Canada's size accounts for seven different time zones. When people in the Yukon Territory sit down to have lunch, people in Newfoundland are getting ready for dinner. If it is 12 noon (1200) in Whitehorse, it is one P.M. (1300) in Vancouver; 2 P.M. (1400) in Calgary; 3 P.M. (1500) in Regina and Winnipeg; 4 P.M. (1600) in Toronto and Montreal; 5 P.M. (1700) in Halifax and 5:30 P.M. (1730) in Newfoundland.

In the chapter "Land–U.S.A." I referred you to the National Institute of Farm and Land Brokers who deal in Canadian as well as U.S. land. There are other private, semiprivate and government agencies in Canada who deal in provincial and territorial land and farms. For Nova Scotia, which has a listing of vacant farmlands available, write: Nova Scotia Farm Loan Board, Department of Agriculture, Truro, Nova Scotia. For private dealers in that province, write: Halifax-Dartmouth Real Estate Board Limited, Granville Street, Halifax, Nova Scotia.

For the Northwest Territories of British Columbia and the Yukon write: Department of Indian Affairs and Northern Development, Box 1500, Yellowknife, N.W.T., Canada, or the Department of Immigration and Manpower, Ottawa, Canada.

5

Developing Uncleared Land

Needless to say, uncleared land can be purchased in the United States and in Canada at greatly reduced prices from that of cleared land or farmland with buildings and livestock already on hand.

However, clearing land and constructing buildings will increase the price of that land no matter how you cut it. If you are young and healthy, clearing your own land with chain (power) saw and ax, mattock, and stump puller, horses or tractor, then using the forest products for the buildings you require (trading timber-sized logs for sawed lumber at your local sawmill or using logs—log cabin style), may be the best way to keep down costs where capital investment is the main barrier to success or failure. I recommend this method only to those of you who are very short of bread. If you have the needed capital to pay for land clearing, I recommend you do so.

If you are in the latter group, take note of the following suggestions:

Before you invest your cash or credit in land-clearing operations, investigate carefully the company whose specialty this is. Check it out with your local County Agent, the closest Agricultural Extension of your State University or, in the case of Canada, with your local forestry land agent. Make it clear to the agent involved you are *not* a commercial farmer and do not want to destroy your land through careless cutting of timber, disregard of springs and streams, indiscriminate destruction of trees which you can use for fuel, and the burning of stumps that can, when used as boundary lines along with barbed-wire fences, furnish refuge for small wild creatures which would otherwise be forced to seek sanctuary elsewhere. Do not allow the destruction of hard maple groves: Maple syrup, sugar and its by-products represent food plus the year's first cash crop.

However little or much land you homestead, purchase or lease, leave a minimum of one half in standing timber, maple groves and second-growth timber (for fuel).

Clearing your own land is hard, slow labor (requiring knowledge and some skill in the use of the tools required), but also highly satisfying and, in the long run, far less wasteful in terms of the timber, maple trees, salvaging of springs and streams and the accumulation of enough second-growth timber for fuel to last you many winters (according to the amount of land you clear). Contrary to the advice given by most experts and authorities on agriculture, I heartily recommend it. While it is underway, set up a campsite or build an Adirondack-style shelter, which is far sturdier than canvas tenting and will see you through a reasonably temperate winter without hardship.

Clearing Priorities

1. The first area to be cleared should be your campsite plus approximately one-half to one acre for your kitchen garden. Two different types of brush cutters available are illustrated.

2. Next to be cleared should be the additional land surrounding the homestead which you will eventually build for permanent residence plus enough land for livestock shelter.

3. Finally to be cleared are the pastureland, hay land and crop land for the growing of animal feed and/or grazing land.

Once a garden site has been selected, stake out a rectangular plot 60 by 90 feet. This area, which has to be cleared of brush, trees and tree stumps, plowed, harrowed and planted, will produce all the fresh and canned vegetables you can eat. Because the farm site we are discussing is forested and inhabited by a variety of game, you should cut a 50-foot swath of cleared land beyond the garden's actual borders. The trees must be felled and hauled away from the immediate area and the brush cut away close to the ground surface. It is not necessary, however, to pull or blast out the tree stumps within this outer swath. To protect your garden from wildlife freeloaders, erect a 6- or 8-foot fence of heavy woven wire around the entire garden. You can use home-grown wooden posts or store-bought steel stakes and further discourage wildlife invasion by festooning the fence with strips of cloth that will flutter in the wind and kerosene lanterns that will burn throughout the hours of darkness.

There are two ways to clear your kitchen garden and create that 50-feet-wide swath of cleared land beyond it: the easy way and the hard way. The easy way is to hire the services of a neighboring farmer and his tractor. If you make your arrangements with him well before he starts his spring planting, the chances are good that you can get him to do the job at a reasonable price. The hard way is to do it yourself, but if you use a chain saw for felling the trees and larger brush, a stump puller for the smaller stumps, dynamite for the large stumps and a brush hook for the lighter brush, you won't find the job too difficult once your muscles become accustomed to the labor.

How to Fell a Tree

If you're going to live on a farm you had better learn how to cut down a tree and do it right—without killing or injuring yourself in the process. The tools you will need include a good chain saw, ax, two steel wedges and a sledge hammer, a peavey to handle cut timber and a horse or tractor, borrowed, rented or purchased to haul the cut timber and stumps clear of the cutting area.

A hand saw for cutting timber can be purchased almost anywhere for $10 or less, but the best investment you can make is in a heavy-duty, fuel-operated power chain saw with a 21-inch bar blade that will fell a tree up to 41 inches through the trunk within a matter of minutes. My personal favorite is a Sears Roebuck "Powersharp" that carries a price tag of only $238, but illustrated (straight from the Sears catalog) are some of the many other saws that

Utility gas-powered Chain Saws
Direct-drive engines deliver plenty of power for general purpose cutting around the house and at the campsite
TOTAL TESTING
Your assurance of quality, dependability . . (see story on preceding page)
SEARS
Deluxe WOODCUTTER
High-performance engine delivers high-speed cutting power
$137.95
Economy WOODCUTTER
Standard-duty engine
$99.95

will perform almost as well. I recommend a heavy-duty saw with a 21-inch bar blade for the tall timber; if your woodlot consists of second-growth stuff, you can get by handily with a lighter-weight rig for about half the above price.

The right method of felling a tree remains the same whether you use a chain saw or hand saw. First you must determine which way the tree is going to fall. If the trunk leans in a particular direction or has most of its heavier branches on one side, that's the side where it's going to fall. If the trunk is true (straight) and the foliage evenly distributed, you can make it fall where you will—providing wind is not a factor.

Mark the trunk in the direction you want it to fall. Hold your saw about waist high and make a straight cut about one-quarter of the way into the trunk. Now, notch the cut. If you are using a power saw, you can make the notch by starting another cut, about 6 inches above the first (according to the thickness of the tree), and sawing down at an angle until you've met the first cut at its furthest penetration point. If you are using a hand saw, the notch will have to be cut out with an ax, using the same technique but a bit more muscle. The tree will fall in the direction of the notch.

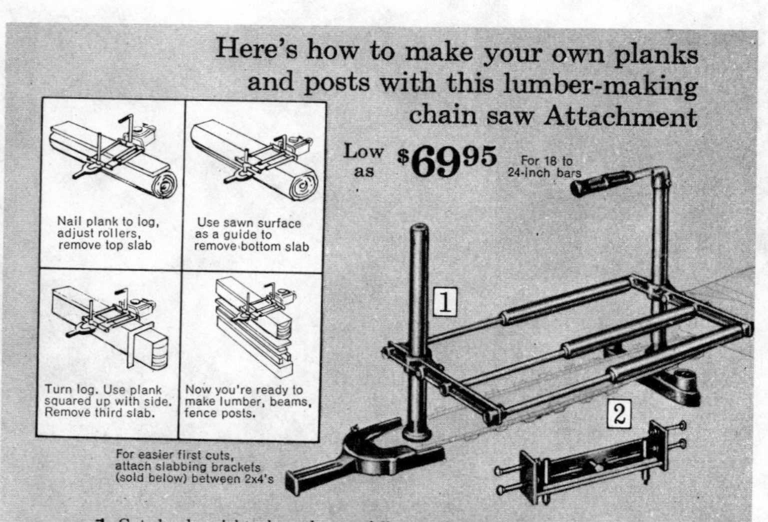

Once the preliminary notch has been cut, start from the opposite side of the tree, making the cut approximately 3 inches above the notch. After you have proceeded through the trunk about two-thirds of the way toward the first cut, the tree may settle and pinch the saw. Drive two steel wedges just deep enough into the new cut to create a "lift" and ease the pinch. *Do not drive them deeper than necessary* because this may cause the tree to topple before it has been cut completely through.

As you cut further and further into the trunk, keep your ears tuned to any sharp *crack* or sudden swaying movement of the tree. Any such sound or movement is a danger signal that the tree is ready to fall. Withdraw wedges and saw and take cover. Should the tree remain standing, even when it has been sawed completely through, drive the wedges deeper into the cut, tapping first one wedge and then the other, until you can see the tree begin to lean. Grab your saw and duck behind the first solid cover you can find.

Trimming

The average to good woodlot consists of a mixture of usable timber, second growth, dead trees, young saplings and brush. Usable timber in the sense applied here means any softwood or hardwood tree 10 or more inches in diameter that has an 8- to 10-foot length of straight trunk. After such a length of timber has been "squared off," it will yield sawn lumber. One way you can do this yourself is illustrated at left.

When clearing ground, trees of timber size should be felled first. Then the second growth and dead stuff, followed by the saplings and brush. Any tree large enough to be used for timber or firewood must be trimmed before being hauled clear of the cutting area. Start by cutting off the top of the tree at the point in the trunk where the diameter is less than 10 inches. Use a saw to trim the large branches, keeping the cut close to the trunk. Smaller branches can be closely trimmed with your ax. Cut the trunk

(Right) *Stump Champion*

into 10- to 16-foot lengths for easy hauling, using the 16-foot figure if the tree is to be cut into lengths for sawn lumber.

Then fell, trim and haul away the second-growth trees and saplings that are thick enough in diameter to furnish you with fuel wood. Use the stump puller shown below for the smaller second-growth stumps.

A brush hook can be handy for cutting the smaller brush. Make several piles of the brush and branches, making sure piles are no closer than 40 to 50 feet apart to avoid fire hazard, then dry for several weeks and burn, using kerosene and/or oil to keep the fire hot. The ashes will make excellent fertilizer for your kitchen garden. Clearing should be done during the winter months or early spring before the maple sap begins to flow.

Pulling Large Tree Stumps

This is where authorities of the Agricultural Extension Services of State Universities and the author part company. Almost to a man these authorities will recommend a power-driven stump champion or chipper similar in design to the one illustrated.

In the manufacturer's instructions you are told to cut the stump as close to the ground as possible and remove the dirt around the stump. The machine will then literally chip the stump out of the ground. I recognize its convenience and feature it here for those who wish to use the device. It is manufactured by the H and H Manufacturing and Supply Co., P.O. Box 345, Greenfield, Mass., under the patent "Stump Champion."

My own preference for the harder job of

pulling the stump, then using the pulled stumps in conjunction with a barbed-wire fence as fencing for property lines and livestock control, has to do with wild-game preservation. Small game cannot live on property cleared of stumps, brush and trees. It needs natural cover such as boundary hedges, stone rows and/or stump rows to protect it from predators. But that is as far as my advice goes; use your judgment for the final decision.

If you decide to pull the stumps and use them as boundary lines, you should proceed as follows:

Cut the tree at least 2½ to 3 feet from the ground to provide leverage for pulling. Dig around the stump to expose the main roots and chop through to separate them from the stump. Use the bulldozer blade of your tractor to loosen the stump, attacking from four directions so as to move the stump from one side to another. Attach a log chain, making sure you fasten it within 3 or 4 inches of the top of the stump. Then use the tractor to pull the stump out. If it is 5 feet in diameter or larger and does not come free from the ground, it will have to be blasted out (in most states by a duly licensed dynamiter) or burned in place.

When the ground is free of stumps, level it and fill the stump holes. Allow the ground to dry out and then plow, harrow or disk. If you purchase a tractor, the tools shown can be bought as attachments. If you rent or borrow a tractor, your neighbor can supply them and show you how to use them. Perhaps he will do the job himself for a negotiated price according to your location.

As for the time of spring planting, this too will vary with location—ranging from March in the south to June 15 in the north and in Canada. For instructions and other information on planting your kitchen garden, see Chapter 13.

Mallet Stump Puller

Another alternative is the "mallet stump puller"—two types are shown here. It can be cheaply made with the tools available on any

Equip your plow to match your own field conditions.

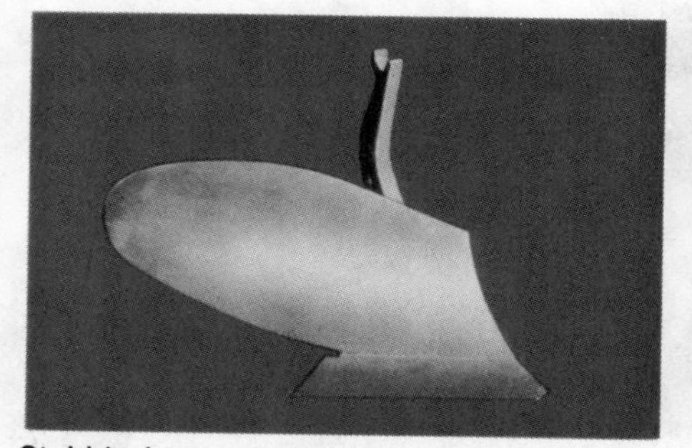

Stubble bottom has wide angle frog which does an aggressive job of turning and pulverizing. Often preferred in hard-to-scour soils. Available in 12, 14, 16 and 18-in. sizes.

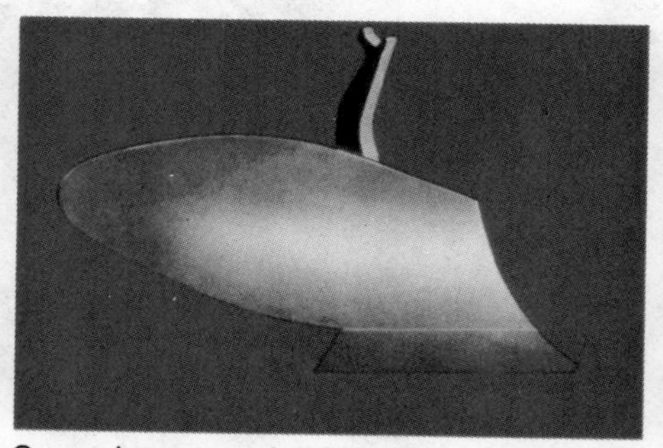

General-purpose bottoms are suitable for practically all conditions. Designed for average plowing. Available in 12, 14 and 16 standard sizes, plus 14 and 16-in. heavy-duty.

Sod and clay bottoms invert furrow slice gently with little pulverizing. Lighter draft and higher speed—works well in heavy sod or clay. Available in 12 and 14-in. sizes.

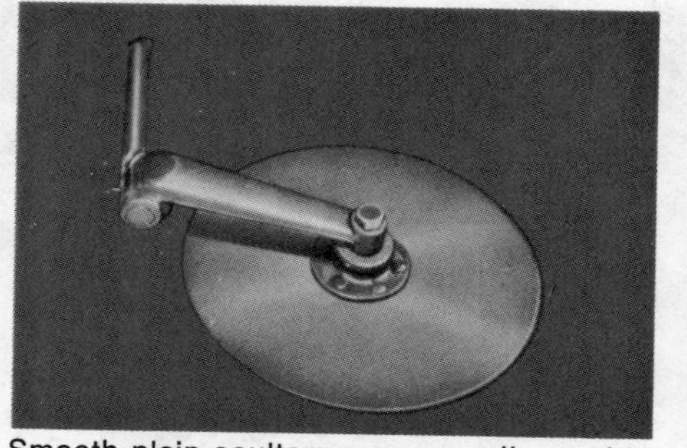

Smooth plain coulters are generally used in sod or relatively clean fields. Single support arm reduces plugging. 1½-in. stems increase strength.

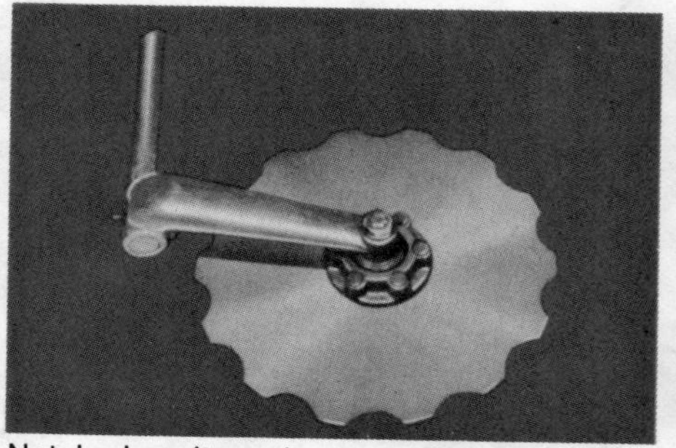

Notched coulters give excellent performance in light to moderate trash conditions. (16 and 18-in. with gray iron or anti-friction bearings.) Wide range of adjustments.

Fluted coulters slash through heavy trash. Fluted design improves traction to keep coulter turning. (17 and 20-in. with gray iron or anti-friction bearings.)

farm. The large mallet head is of cedar or other light wood 18 inches in diameter and 30 inches long; the handle is of hickory or other tough wood about 6 inches in diameter and 6 feet long. A pull chain 12 feet long is fastened to the end of the handle. It is operated by placing the side of the mallet head against the stump with the handle in an upright position. A heavy chain with a "corner bind" or "fid hook" attachment is placed around the stump as low as possible and then around the handle where it enters the log. It is essential that this chain be as tight as possible, since the greatest leverage is obtained when there is no slack.

Pulling the handle down to a horizontal position turns up the stump on the mallet head. The pulling power of the team is increased about six times with this device. Very little strain comes on the mallet head, so the lightest obtainable wood should be used for it, but the handle should be made of tough, strong wood. A 6-inch hole can be made in the mallet head by cutting or burning, but the best way is to bore several small holes within the area of the proposed hole and then chip out the walls between the small holes with a chisel. Good additions but not essential are an iron bolt through the mallet head to keep the handle from coming out, an eyebolt at the top of the handle, and some sheet metal at the lower end of the handle to prevent the chain from wearing into it.

Driving a Well

People who own rural and country homes install their own primary or secondary water-supply systems. If the soil formations permit, driving a well is relatively easy. But to avoid frustration or disappointment, it is wise to check with your State Geological Survey office before starting. If you submit a legal description (survey) of your property, they will advise you whether the conditions in your area are suitable for a well.

It is important to locate a well away from any source of contamination such as marshy areas, cisterns, septic tanks and the like. The well

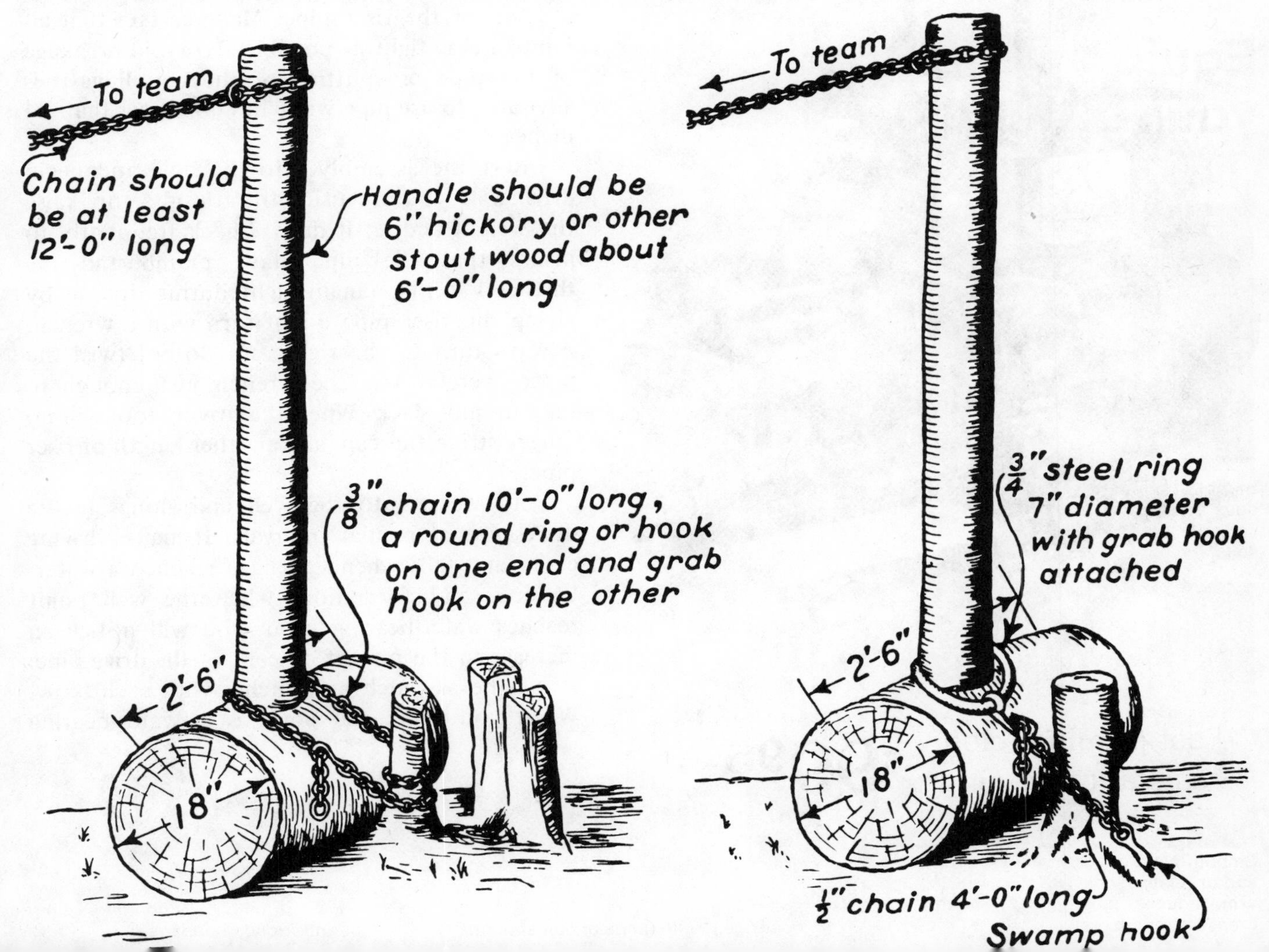

should be situated on higher ground than any of these areas.

Use the best possible equipment. You will need a pump, a well point (1¼-inch diameter recommended),* lengths of galvanized steel riser pipe, couplings, a drive cap and pipe-thread compound. Tools include an auger or a posthole digger, a driving device and pipe wrenches.

A hand-operated pitcher pump is sufficient for driven wells when the water lift does not exceed 22 feet at sea level. At 5,000 feet above sea level the limit of water lift is about 20 feet. Pitcher pumps depend upon a partial vacuum to operate; all joints must be airtight. Pipe compound helps achieve this.

If the water lift in your area is greater than the limits mentioned, a power-driven centrifugal pump (a good cheap one available at Sears is shown here) and 2-inch-diameter equipment must be used. Normally, 40 feet is about the limit to which a 2-inch well can be driven with hand tools. A 2-inch well is not only more difficult to drive than a 1¼-inch well but also requires that a 2-inch drop pipe (with turned couplings) be permanently installed inside.

Driving can be done with a heavy maul or sledge or with a tripod. Square, solid blows are difficult to deliver with the maul and so are not recommended. Glancing blows may break or bend the pipe or strip the threads. Whichever method of driving you decide to use, remember that the riser pipe must be kept perfectly vertical.

The first step is to dig a hole in the ground. The hole can be made with a posthole digger (as shown) or hand auger. Here, again, the hole should be vertical and be dug as deeply as possible to cut down on driving distance.

Assemble the well point, using R & D couplings and pipe-joint compound, to one or more lengths of riser pipe, depending upon the depth of the hole. Fasten a malleable-iron drive cap to the top of the riser pipe. Make certain that all joints are as tight as possible. To avoid breakage of the pipe or splitting of the couplings it is advisable to use pipe wrenches no longer than 24 inches.

Insert the assembly into the hole and begin driving. If you are using the tripod setup, raise the weight and let it drop. Check frequently to insure that the pipe stays plumb and the threaded joints remain tight during driving by giving the riser pipe a half turn with a wrench. Always turn to the right, but do not twist the pipe severely. Use the wrench just enough to take up any slack. When the driving tool will no longer strike the cap, add another length of riser pipe.

Pour water into the well and alongside the drive pipe at regular intervals. It makes driving easier and tells when you have reached a water-bearing sand formation. When the well point reaches water-bearing sand, you will notice an increase in the rate of descent of the drive pipe. It can be as much as 6 inches with each blow. When you think the point is in water-bearing

*A good wellpoint to use is the Red Head brand wellpoint from U.O.P. Johnson Div., 315 N. Pierce St., St. Paul, Minn. 55104.

Fast priming, clog-resistant pumping **$99.95**

sand, pour water into the pipe. If the water stays in the point, you guessed right. If it drains out, it's back to the driver. If the point is in clay, or other nonwater-bearing material, the water will either remain in the pipe or the drop in water level will be extremely slight.

Another method used to check for water is to lower a weighted line into the pipe. When you've hit water, the wet portion of line lets you know how deep the water stands in the well; the dry portion is a measure of depth from the top of the well to water level.

In some instances a greater length of the well point can be brought into contact with water-bearing sand by raising or lowering the tripod assembly about one or 2 feet.

When the well point is at the desired depth, it must be cleaned of sand and muddy water. Cleaning also helps to properly position loose material around the outside of the point, which in turn brings the well up to maximum yield ability. Use either method shown and then remove the fine sand from the well with a pump. Probably the better purging method is to jet water into the well with a garden hose inserted to the bottom of the well. The dirty water and sand will wash up and around the hose. Repeat this flushing procedure until no more sand is obtained by pumping.

Before final installation of the pump, remove all sand particles from its interior, paying particular attention to the valves and plunger. Before drinking any water, ask your State Health Department to test the water for you.

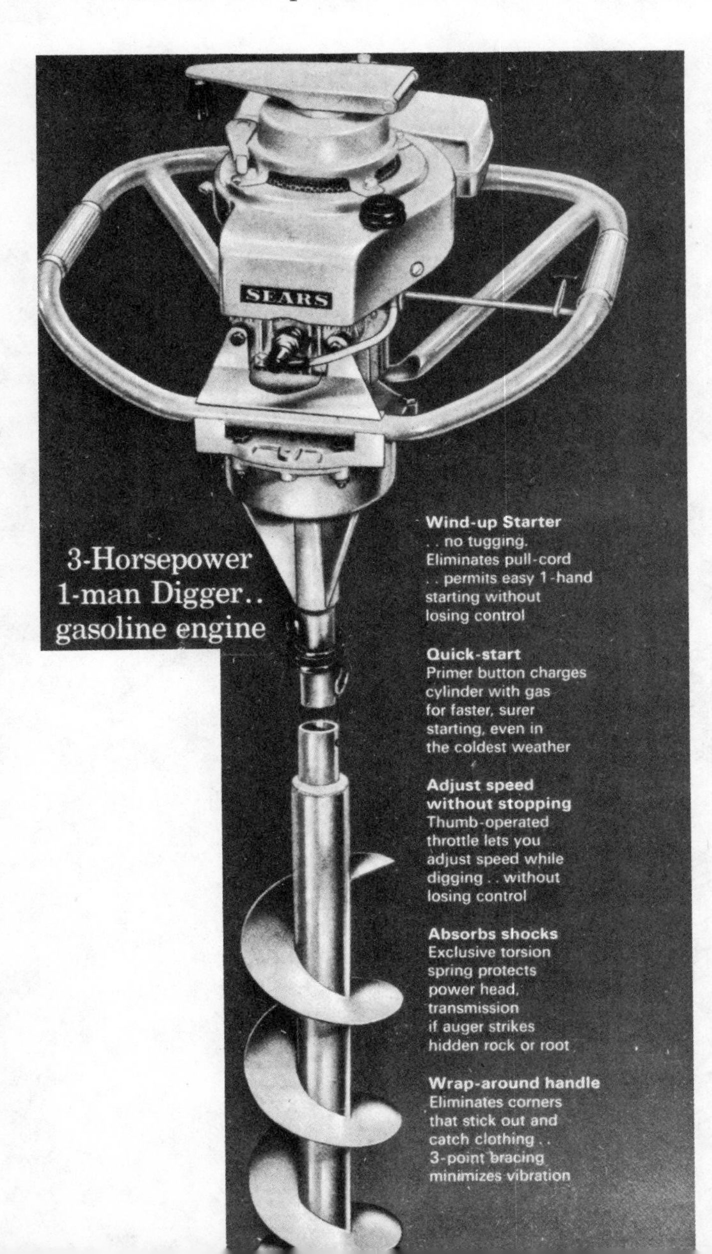

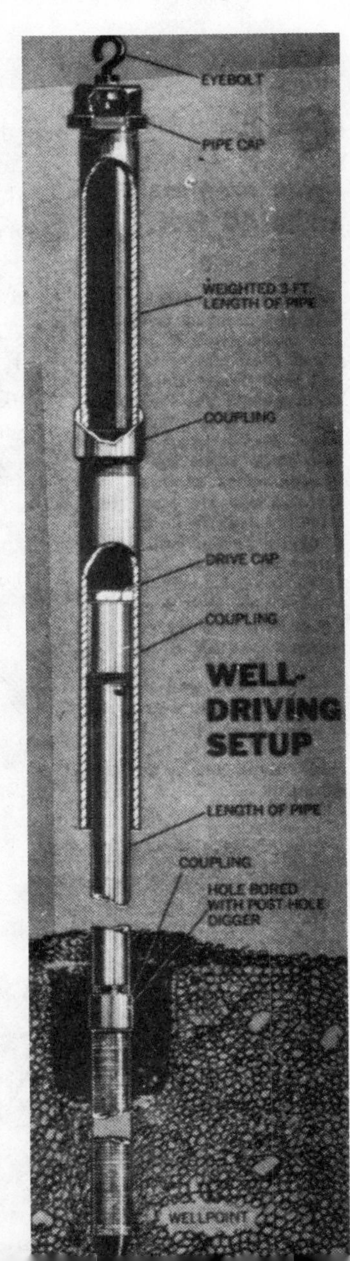

6

Housing

What America has achieved in terms of space travel can also be applied to our more pressing problems on earth. One of the first priorities should be an effort to solve our massive housing problem.

A start has been made. We see the scattered results throughout the country. An outstanding example is in the area of rural housing. The federal government in cooperation with the University of Wisconsin in recent years has developed a program of very low-cost wood housing.

Years of research have developed a home which anyone with the barest knowledge of tools and woodworking may build with his or her own hands. An exploded view of that home is presented here.

Exhaustive step-by-step instructions and 136 clear illustrations can be purchased for $1 from the U.S. Government Printing Office, Washington, D.C., or the Forest Products Laboratory, Forest Service, U.S. Department of Agriculture, University of Wisconsin, Madison, Wisconsin.

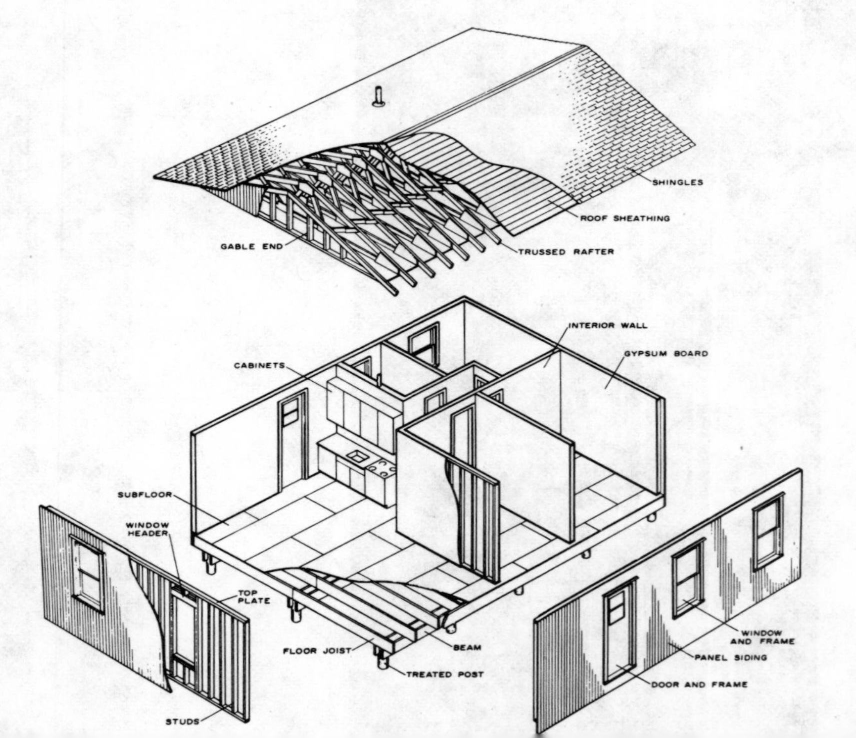

'A' Frame Cabin

Not much more difficult to build but requiring a basic knowledge of blueprints is the 'A' Frame Cabin. Here are the plans.

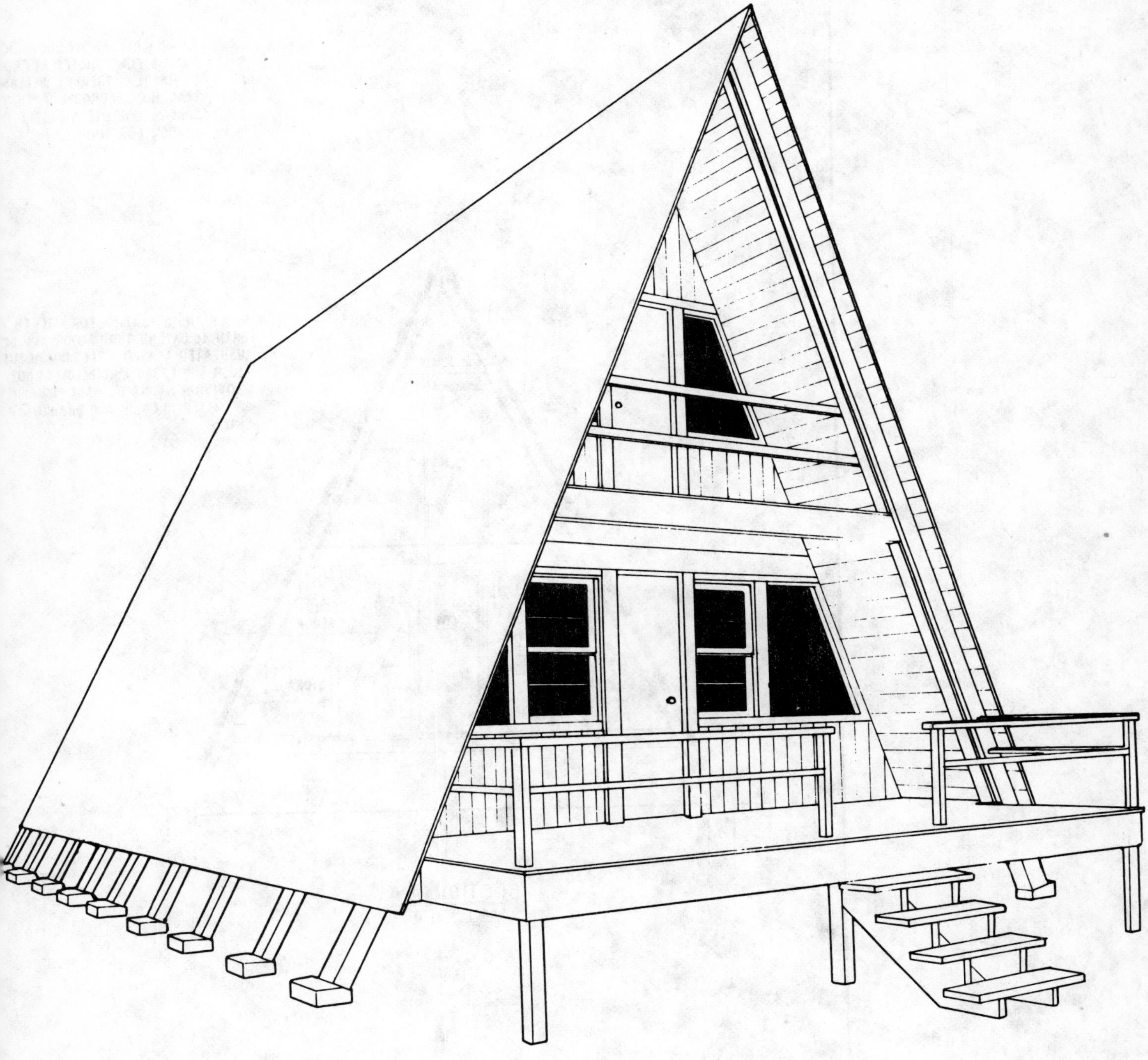

REAR WALL AT SECOND FL
MAY BE CONSTRUCTED AT E
OF RAFTERS THEREBY INCRE
LENGTH OF BEDROOM. IF TH
DONE SUBSTITUTE A DOUBLE
WINDOW FOR DOOR.

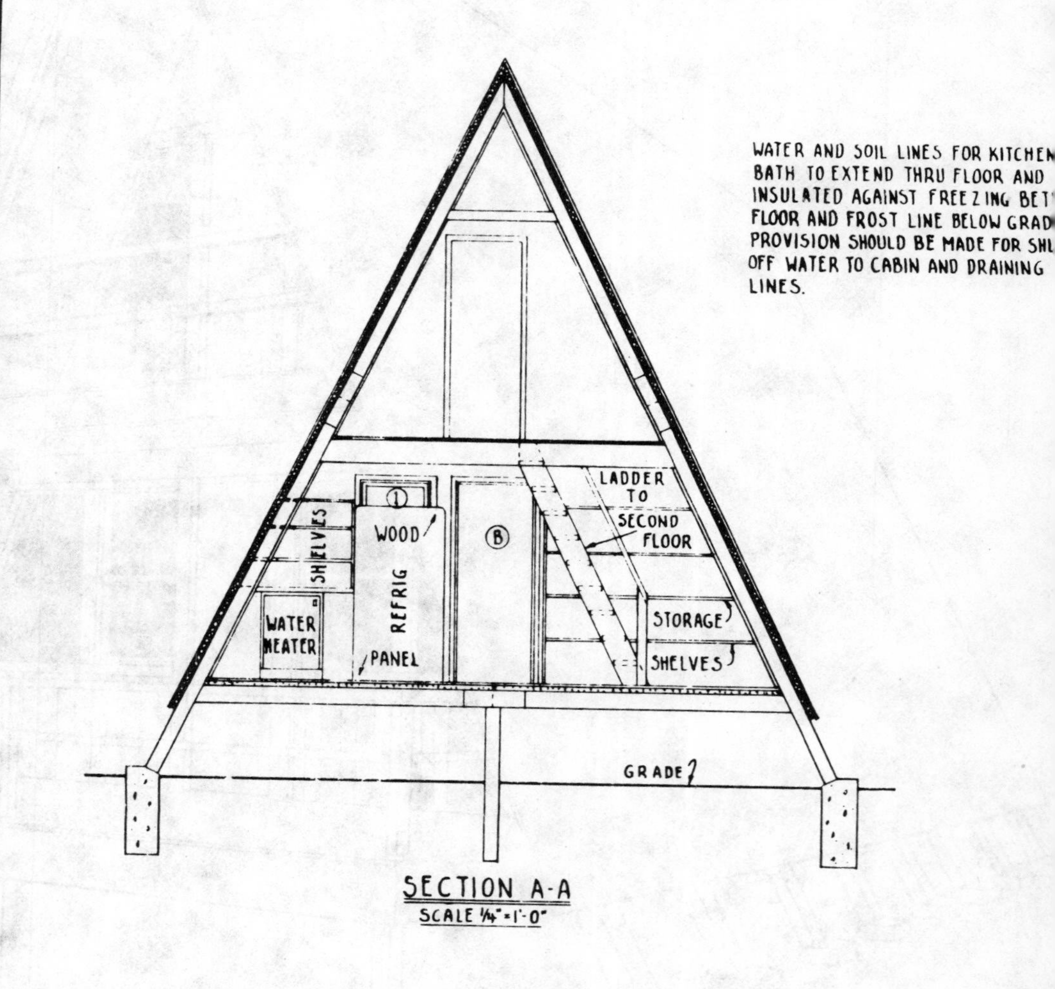

SECTION A-A
SCALE 1/4" = 1'-0"

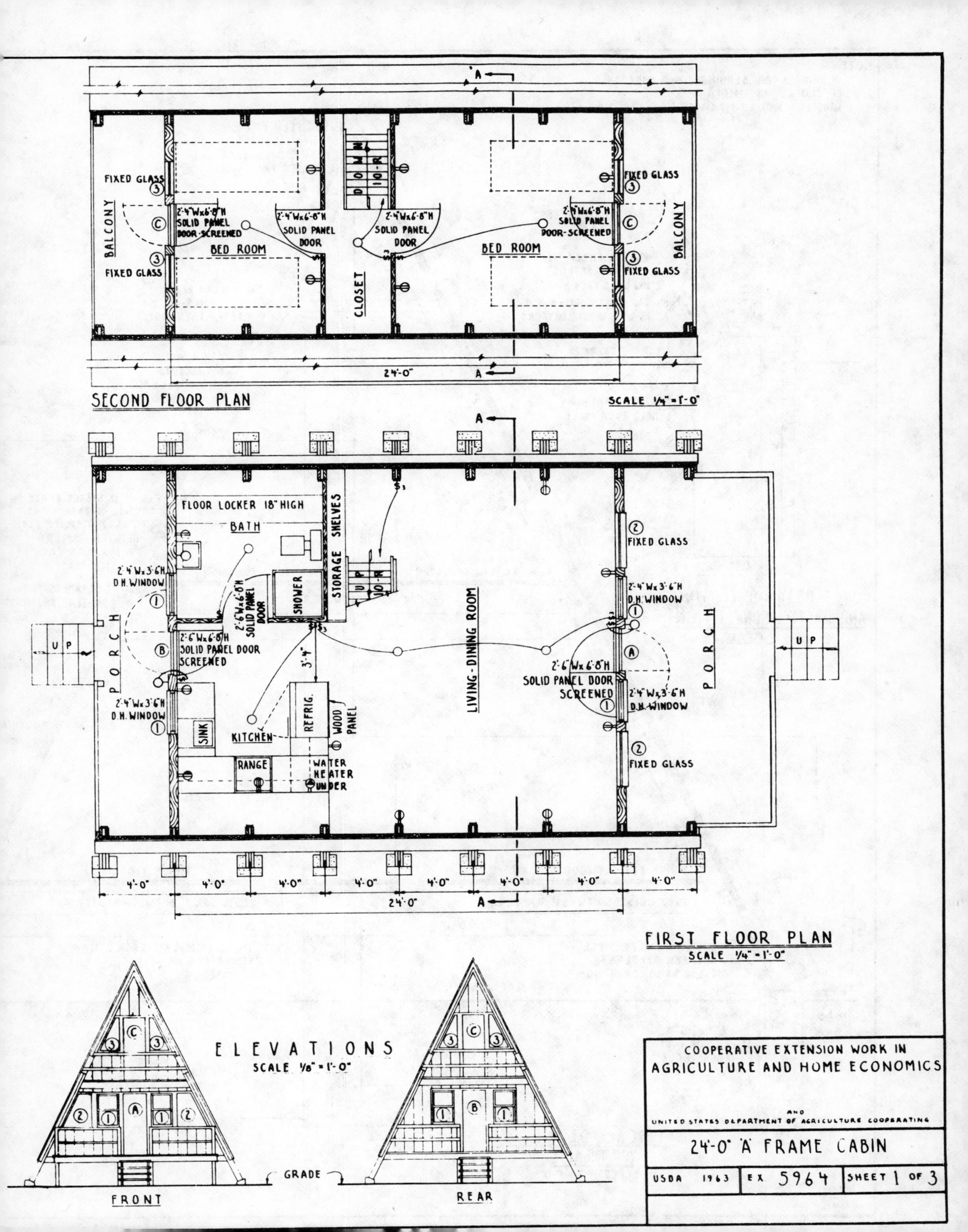
SECOND FLOOR PLAN
SCALE 1/4" = 1'-0"
BALCONY
FIXED GLASS
2'-4" W x 6'-8" H SOLID PANEL DOOR-SCREENED
BED ROOM
2'-4" W x 6'-8" H SOLID PANEL DOOR
DOWN 10-R
CLOSET
24'-0"
FIRST FLOOR PLAN
SCALE 1/4" - 1'-0"
FLOOR LOCKER 18" HIGH
BATH
SHOWER
STORAGE SHELVES
UP 10-R
2'-4" W x 3'-6" H D.H. WINDOW
2'-6" W x 6'-8" H SOLID PANEL DOOR
2'-6" W x 6'-8" H SOLID PANEL DOOR SCREENED
PORCH
UP
KITCHEN
SINK
RANGE
REFRIG.
WOOD PANEL
WATER HEATER UNDER
LIVING-DINING ROOM
FIXED GLASS
4'-0"
24'-0"
ELEVATIONS
SCALE 1/8" = 1'-0"
GRADE
FRONT
REAR
COOPERATIVE EXTENSION WORK IN
AGRICULTURE AND HOME ECONOMICS
AND
UNITED STATES DEPARTMENT OF AGRICULTURE COOPERATING
24'-0" 'A' FRAME CABIN
USDA 1963
EX 5964
SHEET 1 OF 3

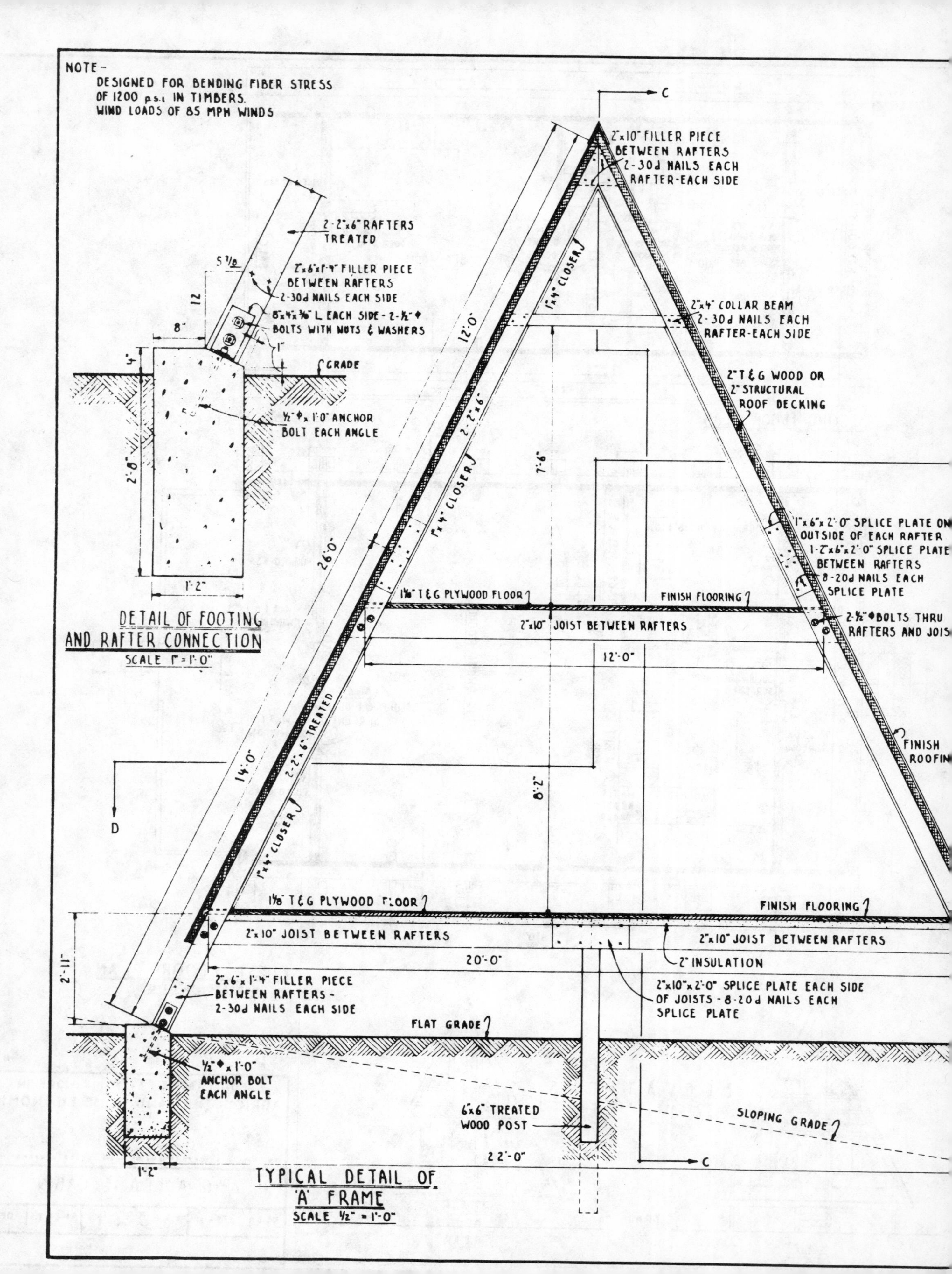

NOTE -
DESIGNED FOR BENDING FIBER STRESS
OF 1200 p.s.i IN TIMBERS.
WIND LOADS OF 85 MPH WINDS
2-2"x6" RAFTERS
TREATED
2"x6"x1'-4" FILLER PIECE
BETWEEN RAFTERS
2-30d NAILS EACH SIDE
8"x4"x3/8" L EACH SIDE - 2-1/2" Φ
BOLTS WITH NUTS & WASHERS
GRADE
1/2" Φ x 1'-0" ANCHOR
BOLT EACH ANGLE
DETAIL OF FOOTING
AND RAFTER CONNECTION
SCALE 1" = 1'-0"
2"x10" FILLER PIECE
BETWEEN RAFTERS
2-30d NAILS EACH
RAFTER-EACH SIDE
2"x4" COLLAR BEAM
2-30d NAILS EACH
RAFTER-EACH SIDE
2" T & G WOOD OR
2" STRUCTURAL
ROOF DECKING
1"x4" CLOSER
2-2"x6"
1"x4" CLOSER
12'-0"
26'-0"
14'-0"
2-2"x6" TREATED
1"x4" CLOSER
7'-6"
8'-2"
1"x6"x2'-0" SPLICE PLATE ON
OUTSIDE OF EACH RAFTER
1-2"x6"x2'-0" SPLICE PLATE
BETWEEN RAFTERS
8-20d NAILS EACH
SPLICE PLATE
1 1/8" T & G PLYWOOD FLOOR
FINISH FLOORING
2"x10" JOIST BETWEEN RAFTERS
2-1/2" Φ BOLTS THRU
RAFTERS AND JOIS
12'-0"
FINISH
ROOFIN
1 1/8" T & G PLYWOOD FLOOR
FINISH FLOORING
2"x10" JOIST BETWEEN RAFTERS
2"x10" JOIST BETWEEN RAFTERS
20'-0"
2" INSULATION
2'-11"
2"x6"x1'-4" FILLER PIECE
BETWEEN RAFTERS -
2-30d NAILS EACH SIDE
2"x10"x2'-0" SPLICE PLATE EACH SIDE
OF JOISTS - 8-20d NAILS EACH
SPLICE PLATE
FLAT GRADE
1/2" Φ x 1'-0"
ANCHOR BOLT
EACH ANGLE
6"x6" TREATED
WOOD POST
SLOPING GRADE
22'-0"
TYPICAL DETAIL OF
'A' FRAME
SCALE 1/2" = 1'-0"

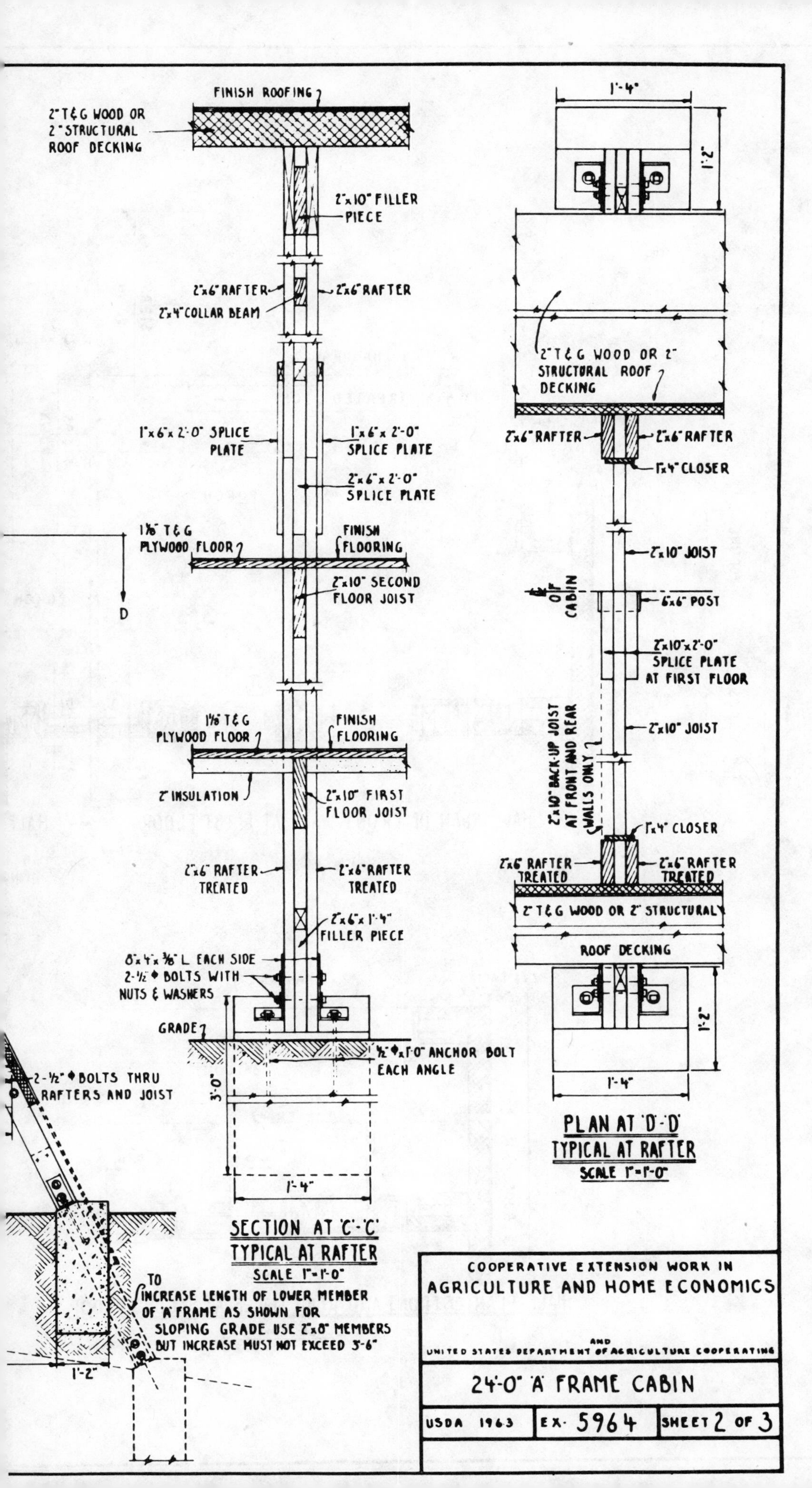
FINISH ROOFING
2" T&G WOOD OR 2" STRUCTURAL ROOF DECKING
2"x10" FILLER PIECE
2"x6" RAFTER
2"x6" RAFTER
2"x4" COLLAR BEAM
1"x6"x 2'-0" SPLICE PLATE
1"x6"x 2'-0" SPLICE PLATE
2"x6"x 2'-0" SPLICE PLATE
1⅛" T&G PLYWOOD FLOOR
FINISH FLOORING
D
2"x10" SECOND FLOOR JOIST
1⅛" T&G PLYWOOD FLOOR
FINISH FLOORING
2" INSULATION
2"x10" FIRST FLOOR JOIST
2"x6" RAFTER TREATED
2"x6" RAFTER TREATED
2"x6"x 1'-4" FILLER PIECE
8"x 4"x 3/8" L EACH SIDE
2-½" ϕ BOLTS WITH NUTS & WASHERS
GRADE
½" ϕ x 1'-0" ANCHOR BOLT EACH ANGLE
3'-0"
1'-4"
2-½" ϕ BOLTS THRU RAFTERS AND JOIST
1'-2"
TO INCREASE LENGTH OF LOWER MEMBER OF 'A' FRAME AS SHOWN FOR SLOPING GRADE USE 2"x8" MEMBERS BUT INCREASE MUST NOT EXCEED 3'-6"
SECTION AT 'C'-'C'
TYPICAL AT RAFTER
SCALE 1"=1'-0"
1'-4"
1'-2"
2" T&G WOOD OR 2" STRUCTURAL ROOF DECKING
2"x6" RAFTER
2"x6" RAFTER
1"x4" CLOSER
2"x10" JOIST
℄ OF CABIN
6"x6" POST
2"x10"x 2'-0" SPLICE PLATE AT FIRST FLOOR
2"x10" JOIST
2"x10" BACK-UP JOIST AT FRONT AND REAR WALLS ONLY
1"x4" CLOSER
2"x6" RAFTER TREATED
2"x6" RAFTER TREATED
2" T&G WOOD OR 2" STRUCTURAL
ROOF DECKING
1'-2"
1'-4"
PLAN AT 'D'-'D'
TYPICAL AT RAFTER
SCALE 1"=1'-0"
COOPERATIVE EXTENSION WORK IN
AGRICULTURE AND HOME ECONOMICS
AND
UNITED STATES DEPARTMENT OF AGRICULTURE COOPERATING
24'-0" 'A' FRAME CABIN
USDA 1963
EX- 5964
SHEET 2 OF 3

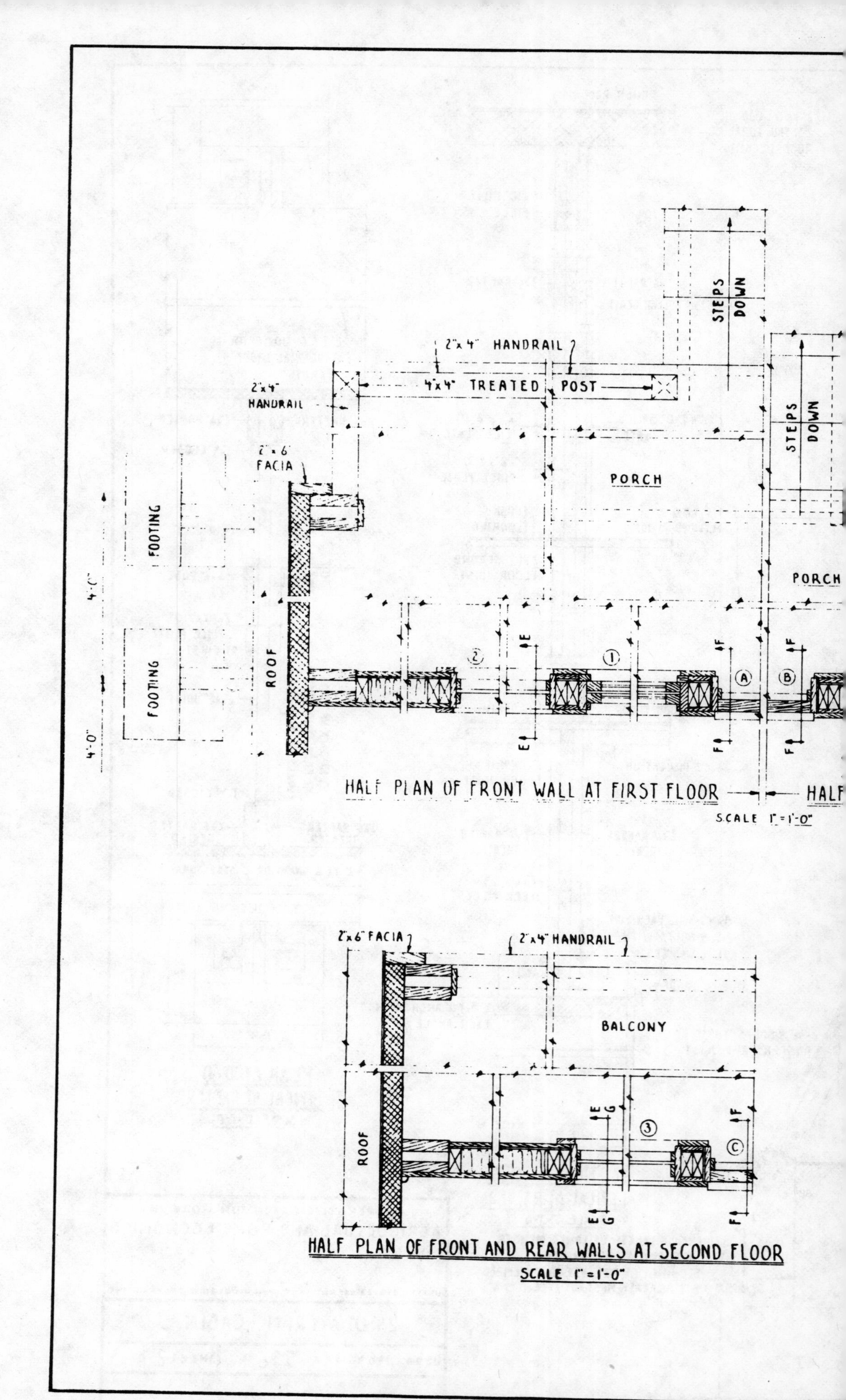

2"x 4" HANDRAIL
2"x4" HANDRAIL
4"x4" TREATED POST
STEPS DOWN
STEPS DOWN
2"x 6" FACIA
PORCH
PORCH
FOOTING
FOOTING
4'-0"
4'-0"
ROOF
HALF PLAN OF FRONT WALL AT FIRST FLOOR
HALF
SCALE 1" = 1'-0"
2"x6" FACIA
2"x4" HANDRAIL
BALCONY
ROOF
HALF PLAN OF FRONT AND REAR WALLS AT SECOND FLOOR
SCALE 1" = 1'-0"

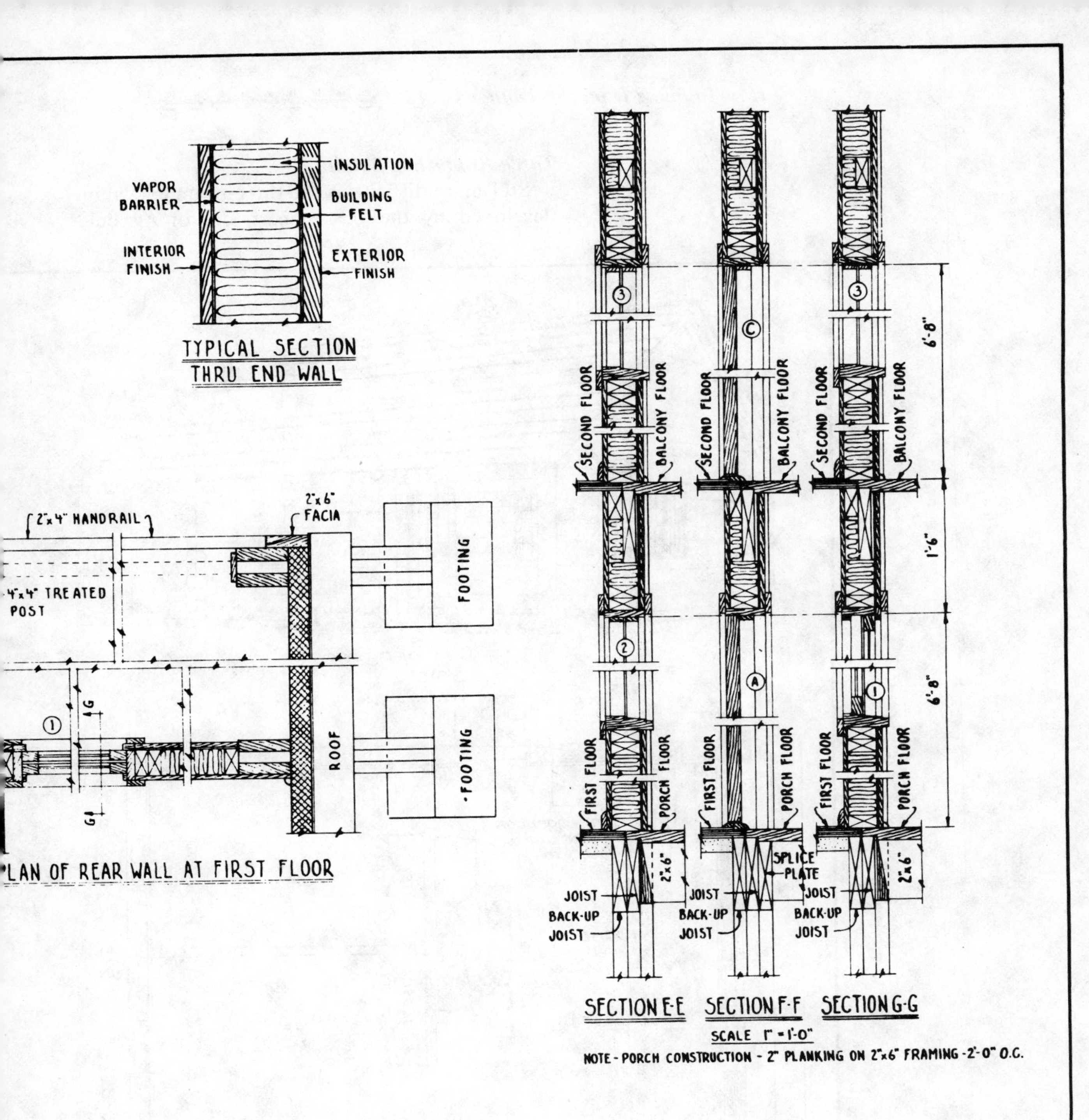

COOPERATIVE EXTENSION WORK IN
AGRICULTURE AND HOME ECONOMICS

AND
UNITED STATES DEPARTMENT OF AGRICULTURE COOPERATING

24'-0" 'A' FRAME CABIN

USDA 1963 | EX. 5964 | SHEET 3 OF 3

Three-Room Log Cabin

Still more difficult is the three-room log cabin developed by the U.S. Department of Agricul-

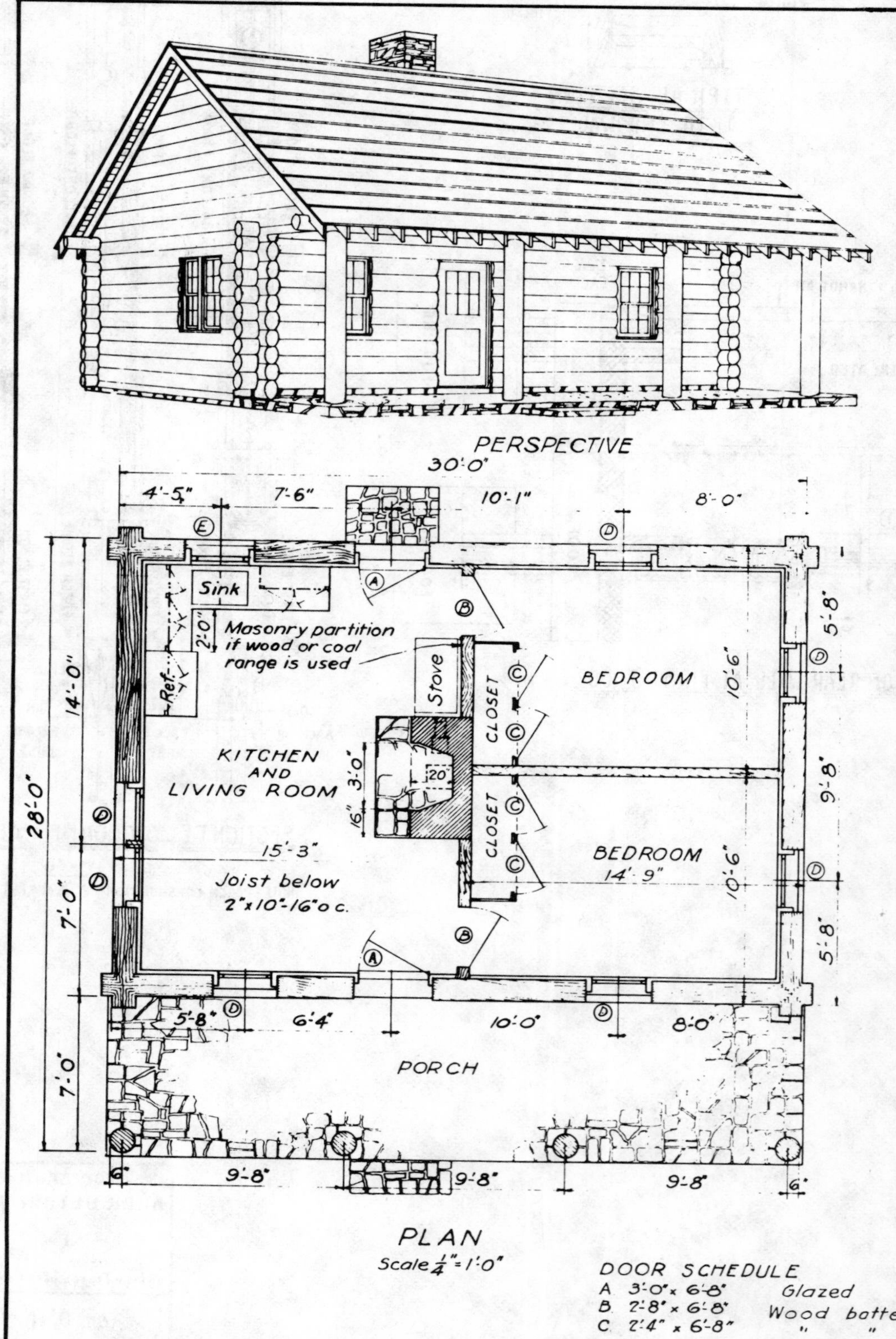

ture in cooperation with the New York State Agricultural Extension of Cornell College, Ithaca, New York.

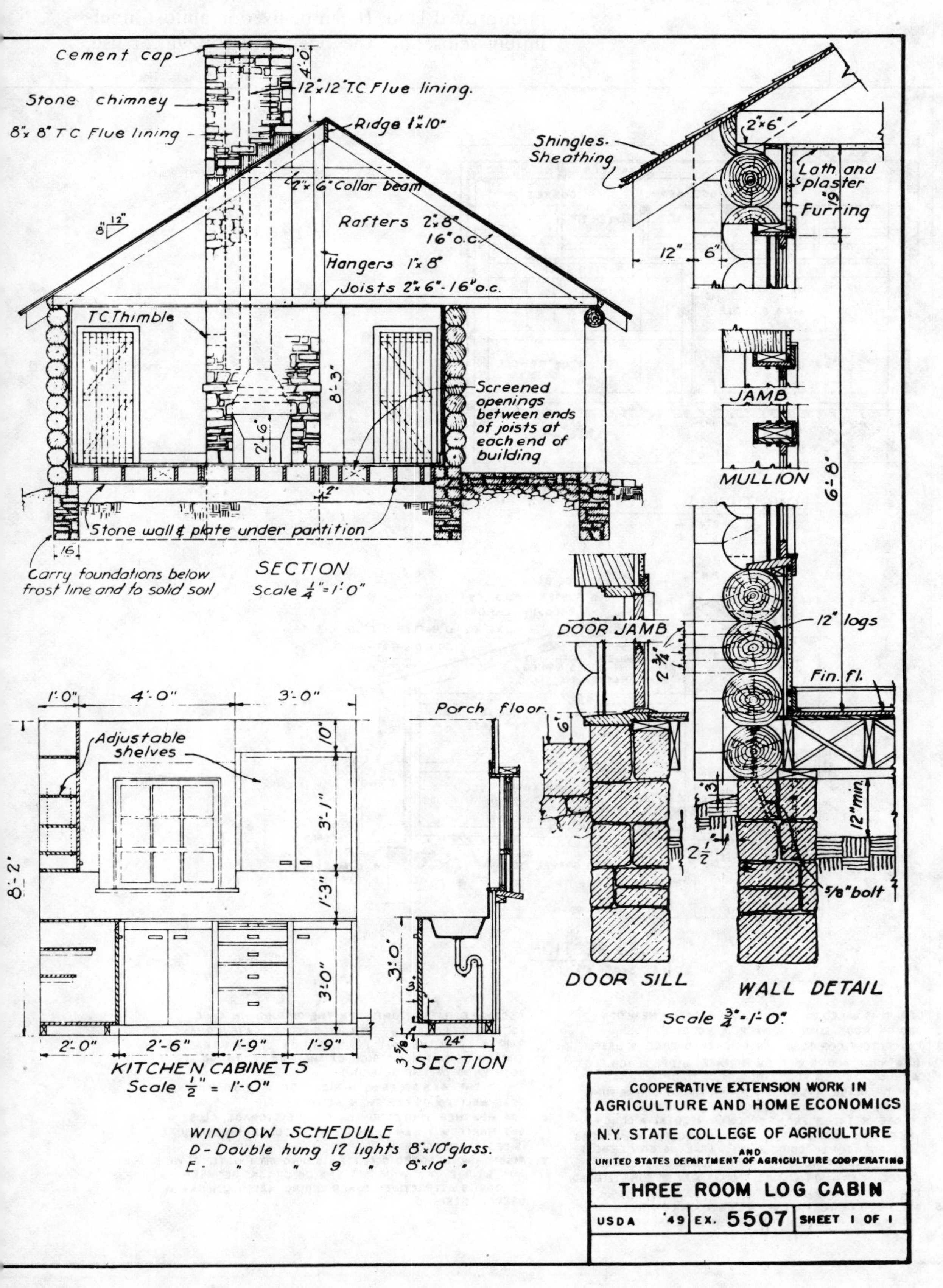

Adirondack-Type Shelter

Simplest of all is the Adirondack Shelter, the first structure you should raise on uncleared, unimproved land. It can be lived in almost indefinitely—close off the open end for winter use,

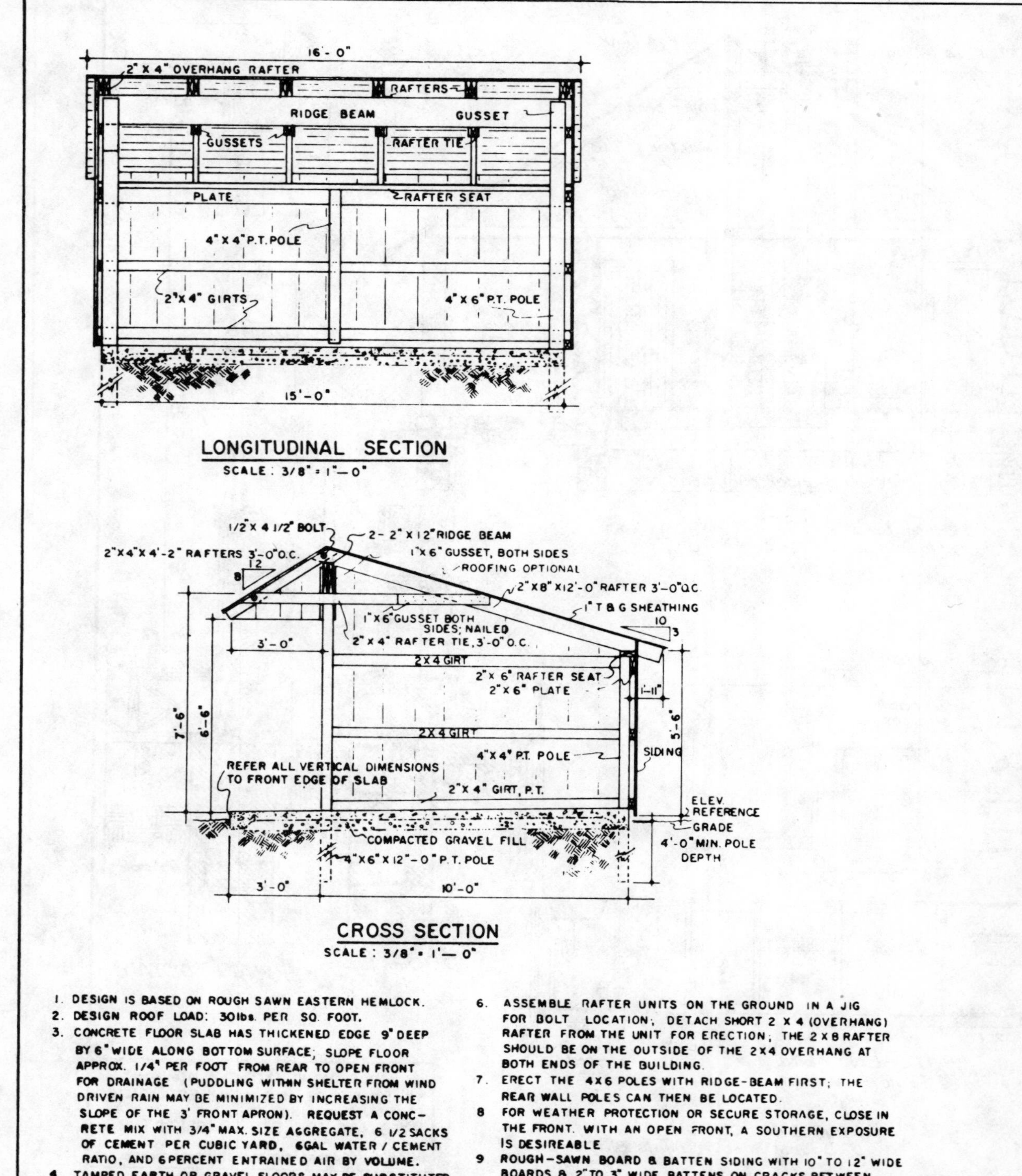

add windows, a door and permanent heating facilities. Later, it may be incorporated into the farm as a machinery shed, a brooder house or other facility.

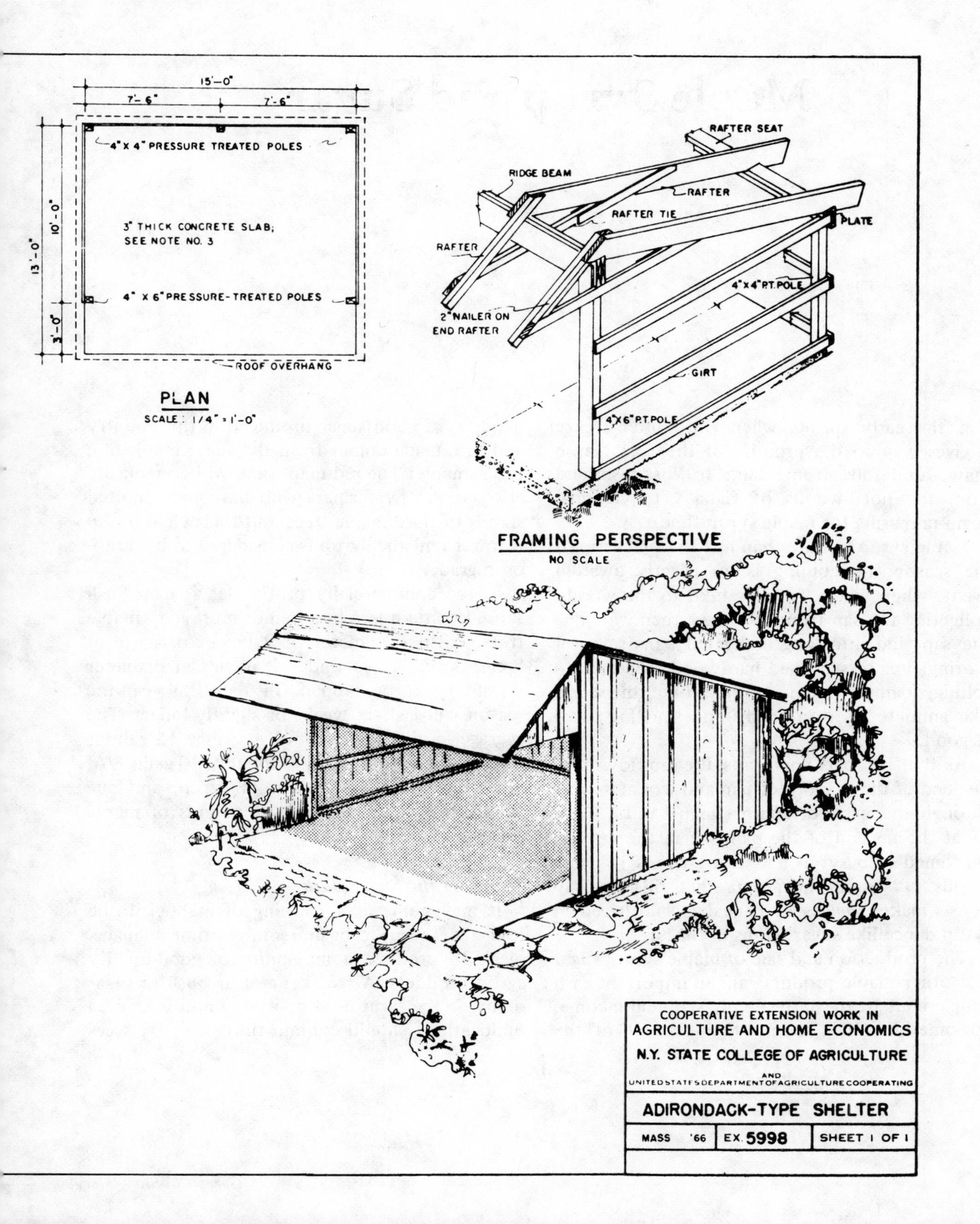

7

Maple Syrup and Sugar

In the early spring, when the winter freeze gives way to the promise of the first March thaw, rural folk, from Maine to Wisconsin and from the northwoods of Canada to Georgia, begin preparing for maple syrup time.

Maple syrup is more than just the first crop of the season; its coming is an eagerly awaited event, when whole families take to the woods, collecting the sap from the trees, then watching the sap, fed into long elongated pans set over roaring fires of seasoned hardwood, boil up into rolling, foaming action, looking for all the world like animated cotton candy and smelling twice as good.

As the fire roars and snaps, the sap flows into one end of the evaporator pan and works its way through the maze of compartments until, finally, at the far end of the pan, where the sap has thickened into syrup and is drawn off in rising clouds of steam, the aroma that fills the sap house makes images of pancakes smothered in syrup dance like sugar plums in the air.

The production and sale of maple syrup, sugar and other maple products are an important cash supplement to the farm family's annual income.

Something like seventy-five percent of the maple syrup and sugar produced in this country and in Canada comes from the sugar maple and black maple. The red maple and white maple are also tapped by farmers who have only limited stands of hard maple trees, but their yield of sap is lower and the syrup is considered to be inferior in grade.

To be commercially profitable, a sugar bush should contain two hundred or more hard maple trees, with a diameter of 9 inches or more at breast level. Trees under 8 inches in diameter should never be tapped. In a typical year the season will last six weeks or slightly longer. The average hard maple tree will produce 15 gallons of sap each season. It takes roughly 40 gallons of sap to produce one gallon of syrup, and one gallon of syrup will yield 8 pounds of maple sugar.

Managing the Sugar Bush

It may come as something of a shock to be told that a sugar bush requires some management to produce a maximum of good-quality syrup products. A well-kept sugar bush or sugar orchard (the term used depends upon your area of location) should contain thirty to fifty trees

Garrett County, Maryland

per acre, depending upon their degree of maturity. A mature tree will measure at least 9 inches in diameter at breast level. If the trees in your sugar bush are crowded, trim them out, letting healthy mature trees stand. They need room to grow and develop spreading crowns; trees with full crowns produce the most sap and prevent too much sunlight from falling on the forest floor. Excessive sunlight will kill the leaf litter and humus, which retain moisture on the forest floor. Maple sap is largely water, and a heavy flow of sap during syrup-making time depends upon soil well supplied with moisture.

Keep livestock out of the maple bush. They'll not only strip leaves and bark off young maple saplings but will reduce the moisture-holding leaf litter and humus.

Cut down soft wood growth, such as pine, hemlock, spruce and fir. Hardwood trees other than maple should also be removed. Some hardwood trees may be left in the maple bush if they are needed to retain a closed crown cover. But as younger maples mature to take their places, these other hardwood trees should be removed.

HARVESTING MAPLE SAP

Prepare Equipment Ahead of Season

The maple syrup season begins with the arrival of the early spring thaws. Most years the thaws come about the middle of March, but they have been known to arrive as early as mid-February. This being the case, it is absolutely essential to have all equipment, from sap buckets and spouts to the evaporator itself, scrubbed and scalded and ready to use well ahead of time. Since the early sap runs give the lightest-colored and best-flavored syrup, the start of the run shouldn't be missed for lack of preparation. Remember that a good syrup season depends entirely upon weather conditions. Prime syrup weather consists of alternate freezing and thawing, with crisp cold nights and mild sunny days.

Equipment

If the idea is to tap only enough trees near the house or farm buildings for home consumption of syrup and sugar, the boiling of syrup may be done in a large kettle on the kitchen stove, or in a cauldron over an open arch built out of doors. Very little additional equipment is needed.

On the other hand, if two hundred or more trees are tapped, your goal will be to produce a commercially profitable crop that will require you to have on hand the following articles of syrup-making equipment.

Tapping Bit: This is a short, strong oval-tipped bit with a coarse-threaded, sharp screw which cuts into the tree rapidly and smoothly. A bit 3/8 inch in diameter is sufficient, since it allows the use of a 7/16-inch reamer. Both bit and reamer will be accommodated by a standard carpenter's brace.

Reamer: If the flow of sap is slowed by a period of unseasonably warm weather, bacterial growth will develop in the tap hole, resulting in sour sap and dark-colored syrup. To correct this condition, ream out the tap holes.

Sap Spouts: Most syrup makers prefer a metal spout which is easily cleaned and will not corrode or rust. The end of the spout that is driven into the tree should be round, smooth and sufficiently tapered to fit bores of varying sizes. It must be strong enough to be driven into the tree with a mallet. Make sure that a hook or notch on which to hang the bucket is provided with the spout.

Buckets and Covers: Although a limited number of wooden buckets are still in use, metal or plastic buckets that are light, easy to clean and can be stored in a limited space are recommended. They should have a capacity of 12 or 13 quarts. A small number of firms still manufacture leaded tin plate or tern-plate buckets, but these should be avoided because of the dangers of lead contamination. Bucket covers to match are essential to prevent sticks, dirt, rain and snow from dropping into the buckets and fouling the sap. Covers also keep the sap cool. They should be ridged or arched in the middle to permit maximum circulation of air.

Gathering Pails: These should be of 18-quart capacity and made of heavy tin plate or galvanized iron. The sap gatherer needs two pails, one in each hand, to balance the sap load.

Gathering Tank: A three-barrel-size gathering tank is just about large enough for a two hundred-tree operation. It comes with a splash cover, a top strainer and an outlet pipe and is mounted on a low sled or "stone boat." Supplement the top strainer with a piece of cheesecloth in order to obtain careful straining of the sap. The tank is drawn by a team of horses or tractor.

Storage Tank: Storage tanks are made of metal and come in capacities of 8, 10, 15 and 20 barrels or greater. Since two hundred average-sized maple trees will accommodate five hundred sap buckets, a 10-barrel storage tank will do for an operation of this size. The tank should be placed on the north or east side of the sap

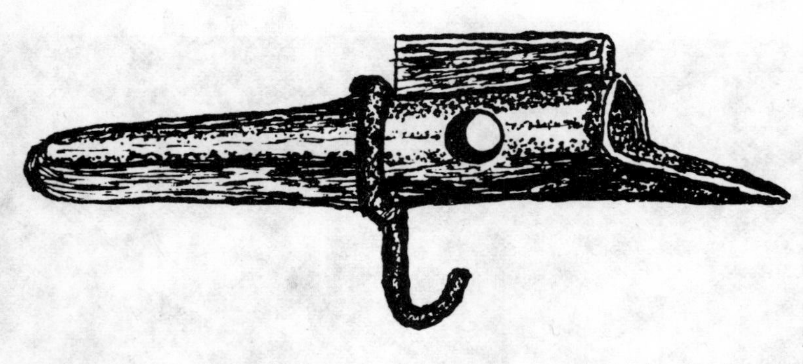

(Above) *Maple sap spout, inserted in maple trees with light taps of hammer. Sap bucket with cover is hung on hook.*

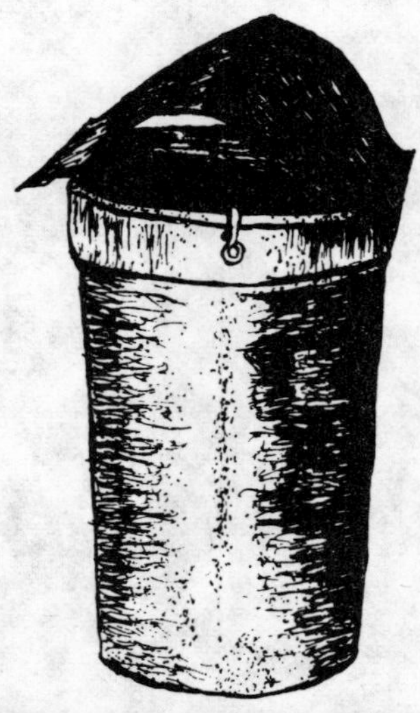

(Left) *Sap bucket with cover is hung on sap spout to collect sap from maple tree.*

Boiling house where syrup and sugar are made. Sap from gathering tank is siphoned off into holding tank behind house (not shown).

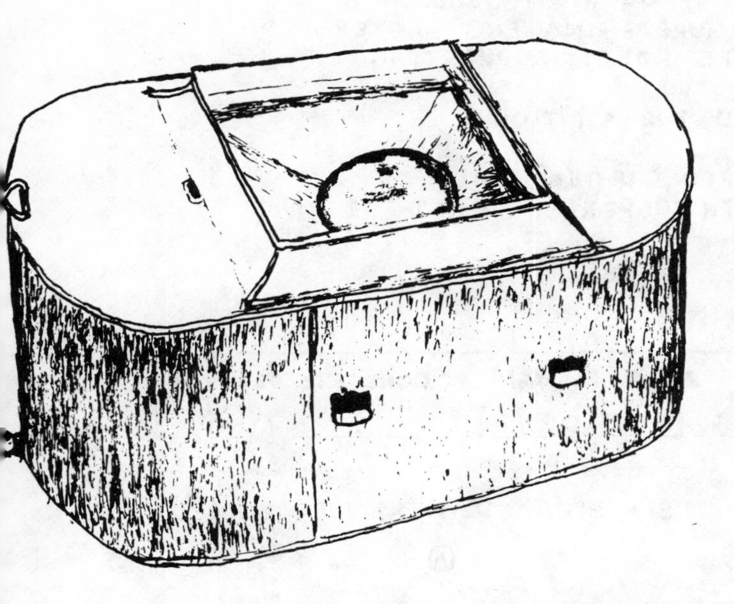

(Above) *Sap is poured from gathering buckets into collecting tank. Tank, which sits on "stone sled" is tractor- or horse-drawn.*

house, out of doors. It must be roofed over and set higher than the evaporator inside, so that the sap may flow into the evaporator by gravity.

Sap House: The sap house, or sugar house, as it is more often called, is the heart of every syrup-making operation. Here is where the evaporator is located that boils down the sap into syrup and sugar. Locate the building close to the maple grove, on the downgrade from the trees to be tapped, or on the same level. If possible, build the structure near a good supply of running water. For a two-hundred-tree operation, a building 20 feet long, 8 feet wide and 7 or 8 feet high at the sides is sufficient. It should be covered with a gable roof, and topped by a ventilator as long as the evaporator, which, in this case, will be 12 feet long. A door at least 40 inches wide should be located at the front end, nearest to the furnace doors of the evaporator. Allow for enough windows to permit plenty of light. The metal chimney should extend well beyond the roofline to ensure enough draft for

NOTE:

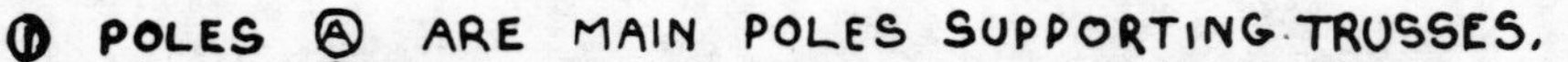

1. POLES Ⓐ ARE MAIN POLES SUPPORTING TRUSSES.
2. POLES Ⓑ ARE MAIN POLES SUPPORTING END RAFTERS.
3. POLES Ⓒ ARE INTERMEDIATE POLES SUPPORTING DOORS AND PARTITIONS
4. EXTEND LENGTH AS REQUIRED FOR ADDITIONAL STORAGE CAPACITY.
5. ALTERNATES: USE RUBBER TIRED CART (NO RAILS), OR OVERHEAD SINGLE MONORAIL WITH SUSPENDED BUCKET FOR WOOD TRANSPORT.

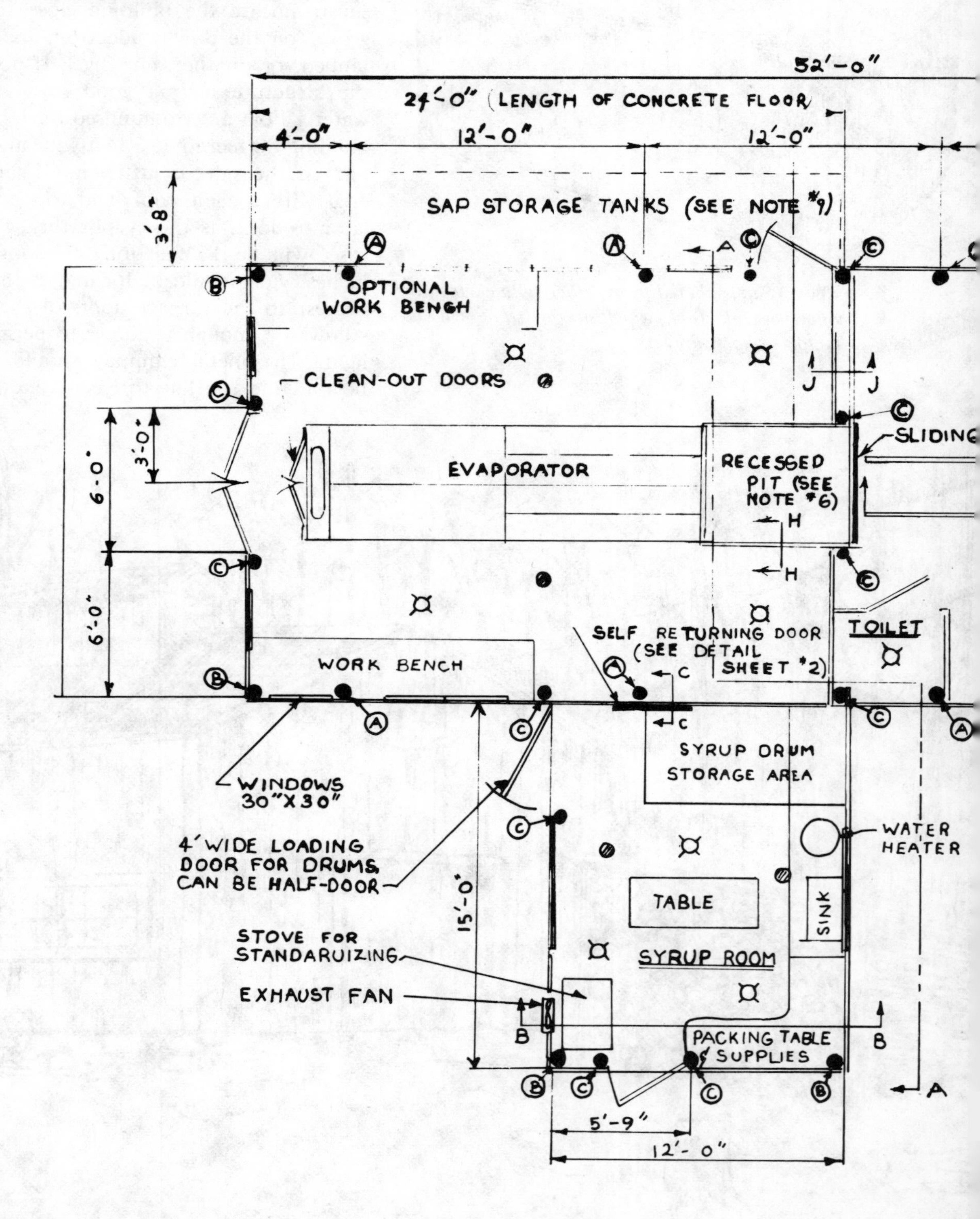

⑥ IF WOOD IS NOT USED AS FUEL ELIMINATE PIT AND CONSTRUCT LEVEL FLOOR ENTIRELY AROUND EVAPORATOR

⑦ ☼ = LAMP LOCATION

⑧ ⊘ = DRAIN LOCATION

⑨ OMIT WINDOWS IN WALL ON THIS SIDE OF BLDG. IF ELEVATED GRAVITY FEED SAP STORAGE TANKS ARE USED INCLUDE WINDOWS IF PUMP IS USED TO EMPTY STORAGE TANKS INSTALLED AT GROUND LEVEL

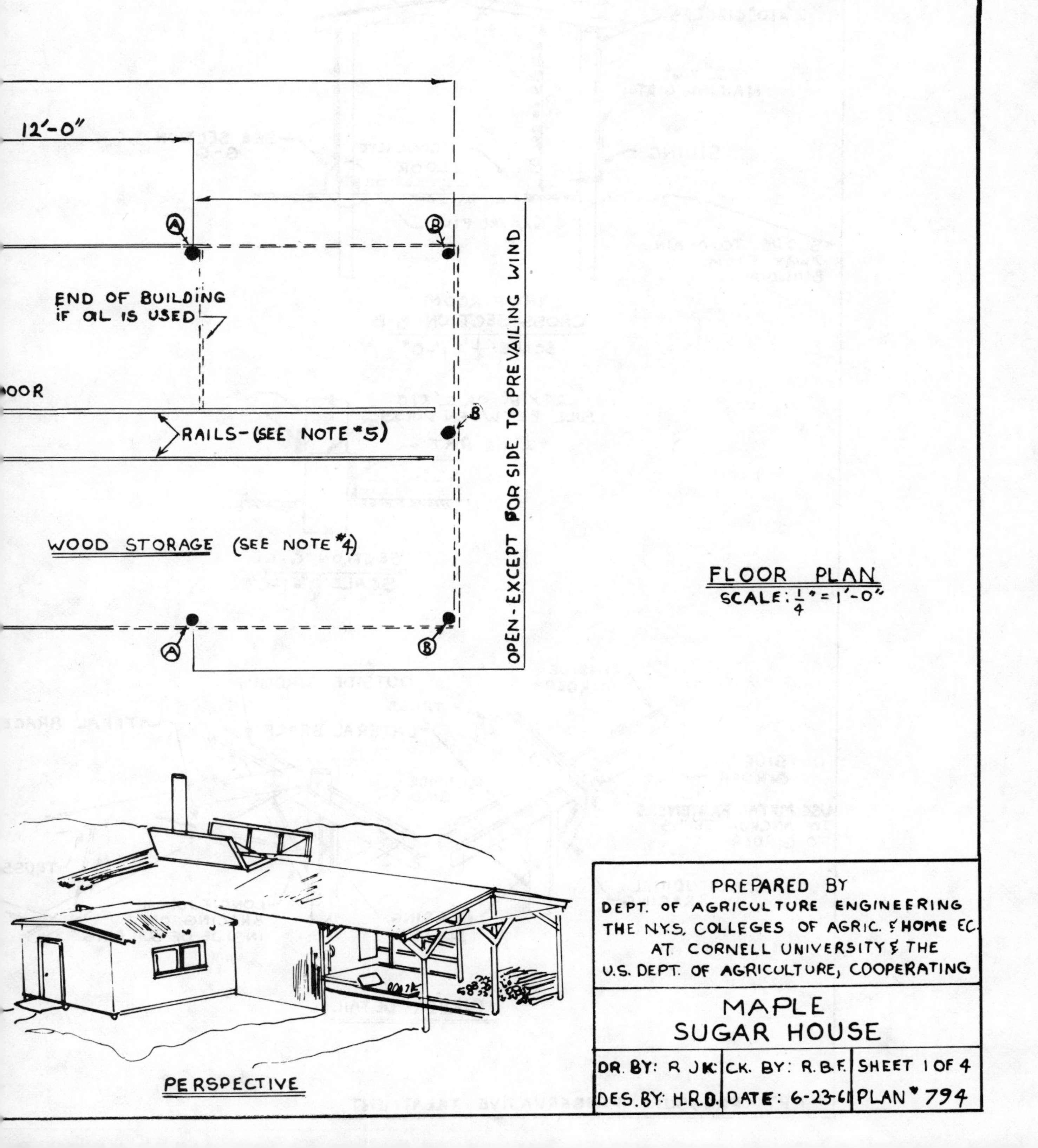

PREPARED BY
DEPT. OF AGRICULTURE ENGINEERING
THE N.Y.S. COLLEGES OF AGRIC. & HOME EC.
AT CORNELL UNIVERSITY & THE
U.S. DEPT. OF AGRICULTURE, COOPERATING

MAPLE
SUGAR HOUSE

DR. BY: R JK	CK. BY: R.B.F.	SHEET 1 OF 4
DES. BY: H.R.D.	DATE: 6-23-61	PLAN # 794

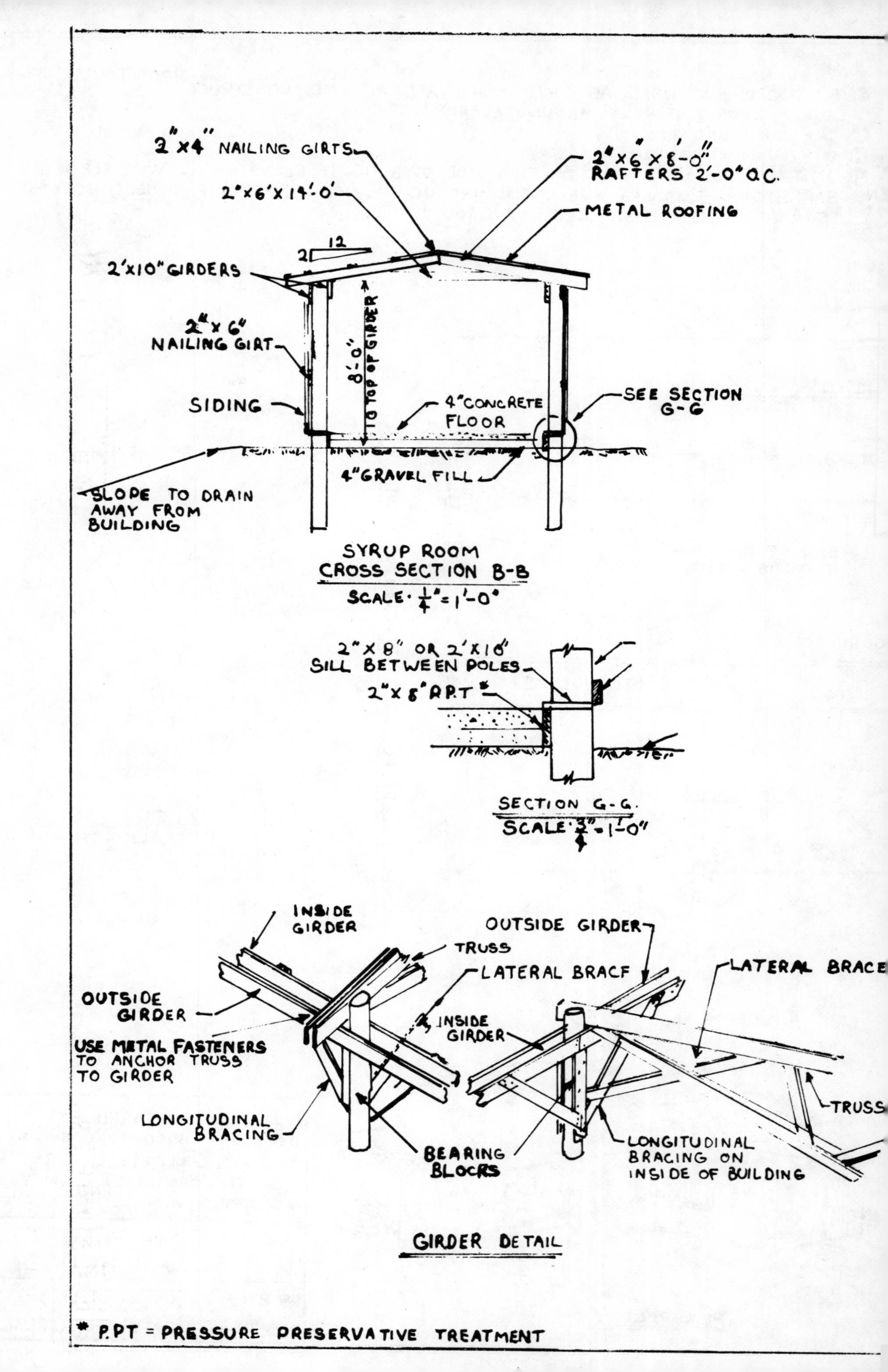

2"x4" NAILING GIRTS
2"x6'x14'-0'
2"x6"x8'-0"
RAFTERS 2'-0" O.C.
METAL ROOFING
12
2
2'x10" GIRDERS
2"x6"
NAILING GIRT
8'-0" TO TOP OF GIRDER
SIDING
4" CONCRETE
FLOOR
SEE SECTION
G-G
4" GRAVEL FILL
SLOPE TO DRAIN
AWAY FROM
BUILDING
SYRUP ROOM
CROSS SECTION B-B
SCALE. 1/4"=1'-0"
2"x8" OR 2'x10"
SILL BETWEEN POLES
2"x8" P.P.T*
SECTION G-G.
SCALE: 3/4"=1'-0"
INSIDE
GIRDER
OUTSIDE GIRDER
TRUSS
LATERAL BRACF
LATERAL BRACE
OUTSIDE
GIRDER
INSIDE
GIRDER
USE METAL FASTENERS
TO ANCHOR TRUSS
TO GIRDER
LONGITUDINAL
BRACING
BEARING
BLOCRS
TRUSS
LONGITUDINAL
BRACING ON
INSIDE OF BUILDING
GIRDER DETAIL
* P.PT = PRESSURE PRESERVATIVE TREATMENT

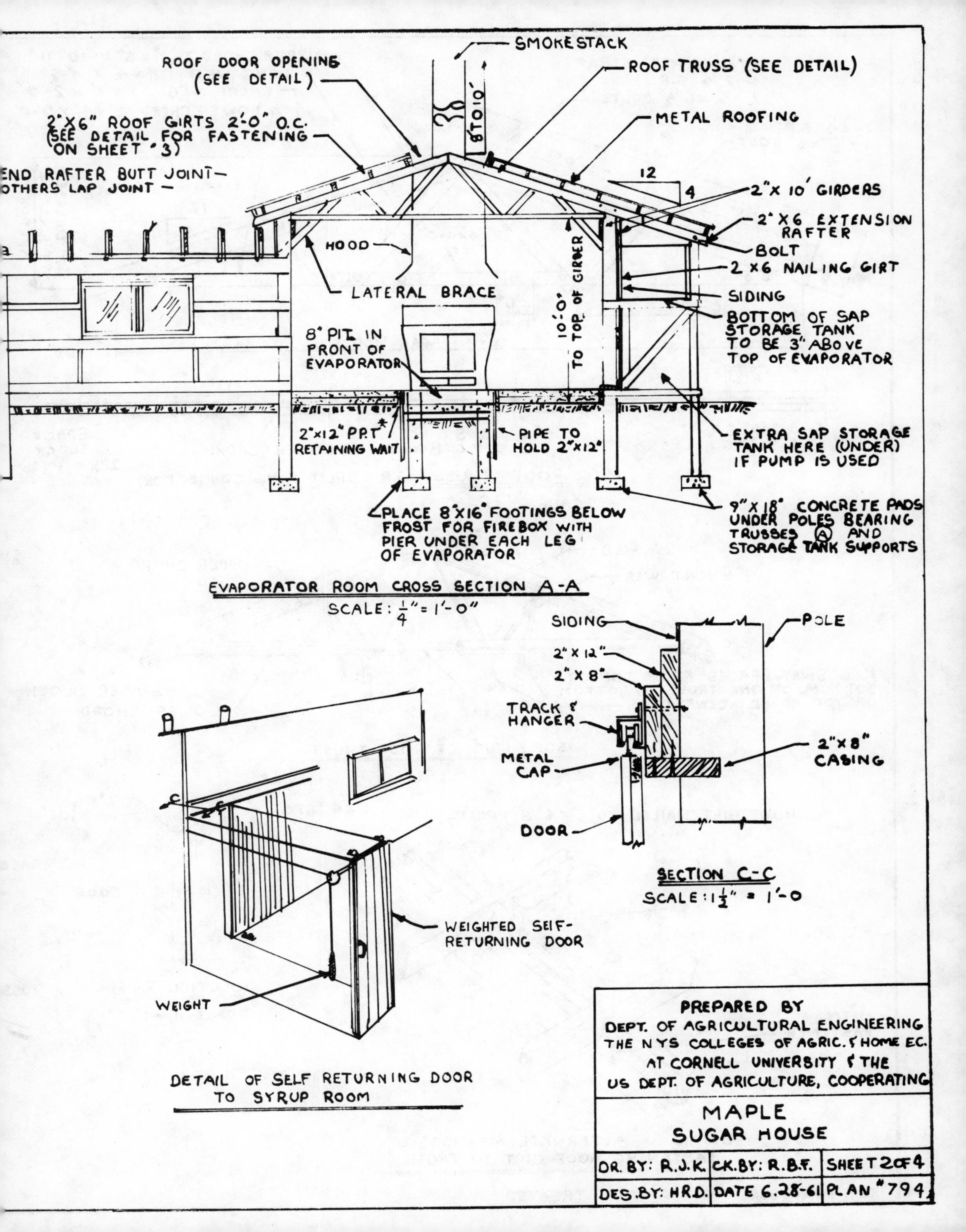
SMOKESTACK
ROOF DOOR OPENING (SEE DETAIL)
ROOF TRUSS (SEE DETAIL)
8' TO 10'
2"X6" ROOF GIRTS 2'-0" O.C. (SEE DETAIL FOR FASTENING ON SHEET 3)
METAL ROOFING
END RAFTER BUTT JOINT—
OTHERS LAP JOINT —
12
4
2"X 10' GIRDERS
2"X6 EXTENSION RAFTER
BOLT
HOOD
2 X6 NAILING GIRT
LATERAL BRACE
SIDING
10'-0" TO TOP OF GIRDER
BOTTOM OF SAP STORAGE TANK TO BE 3" ABOVE TOP OF EVAPORATOR
8" PIT IN FRONT OF EVAPORATOR
2"x12" P.P.T RETAINING WALL
PIPE TO HOLD 2"x12"
EXTRA SAP STORAGE TANK HERE (UNDER) IF PUMP IS USED
PLACE 8"X16" FOOTINGS BELOW FROST FOR FIREBOX WITH PIER UNDER EACH LEG OF EVAPORATOR
9"X18" CONCRETE PADS UNDER POLES BEARING TRUSSES (A) AND STORAGE TANK SUPPORTS
EVAPORATOR ROOM CROSS SECTION A-A
SCALE: 1/4" = 1'-0"
SIDING
POLE
2" X 12"
2" X 8"
TRACK & HANGER
2"X8" CASING
METAL CAP
DOOR
SECTION C-C
SCALE: 1 1/2" = 1'-0
C
C
WEIGHTED SELF-RETURNING DOOR
WEIGHT
DETAIL OF SELF RETURNING DOOR TO SYRUP ROOM
PREPARED BY
DEPT. OF AGRICULTURAL ENGINEERING
THE NYS COLLEGES OF AGRIC. & HOME EC.
AT CORNELL UNIVERSITY & THE
US DEPT. OF AGRICULTURE, COOPERATING
MAPLE
SUGAR HOUSE
DR. BY: R.J.K.
CK.BY: R.B.F.
SHEET 2 OF 4
DES.BY: H.R.D.
DATE 6.28-61
PLAN #794

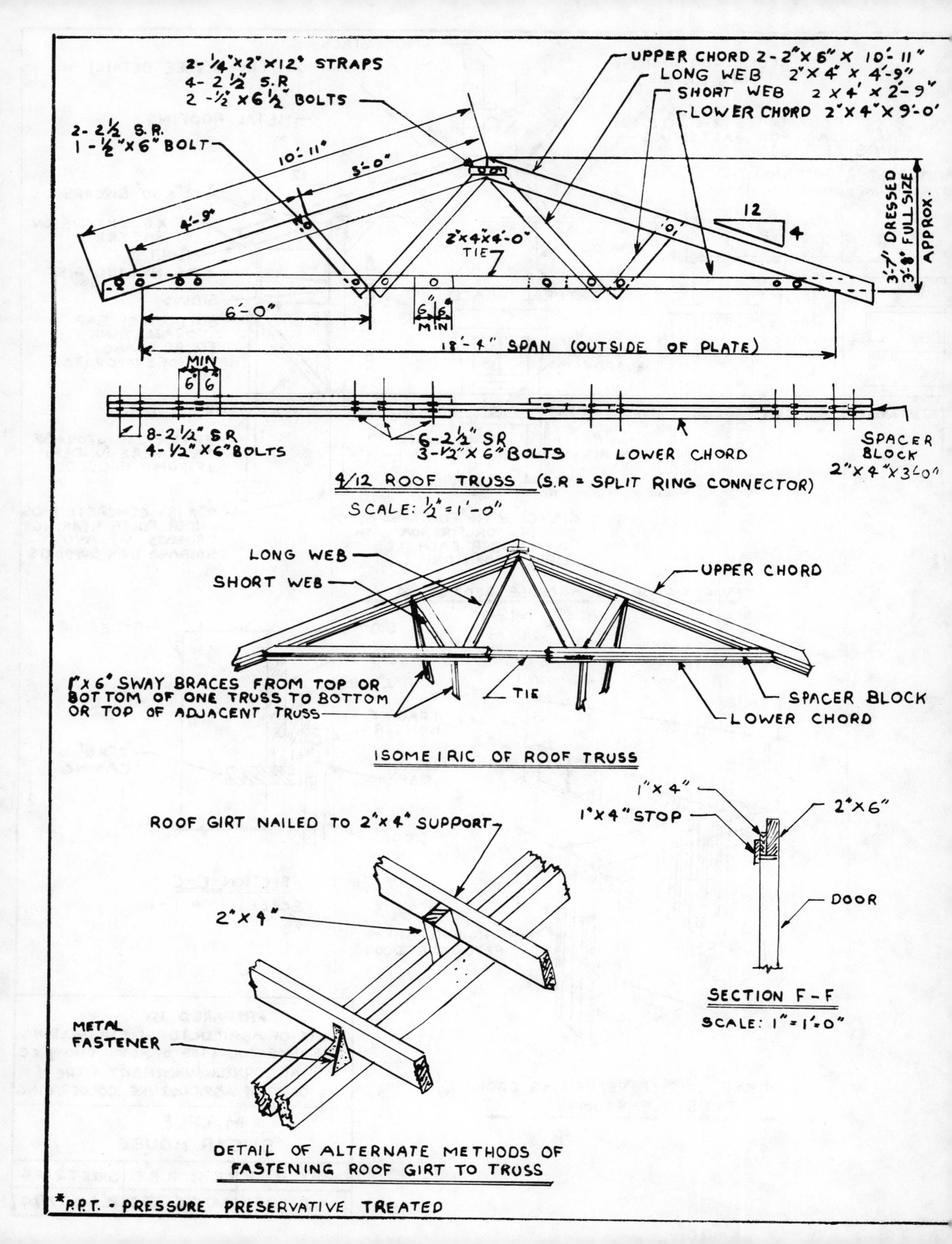

2-1/4"x2"x12" STRAPS
4-2 1/2" S.R
2-1/2"x6 1/2" BOLTS
2-2 1/2 S.R.
1-1/2"x6" BOLT
UPPER CHORD 2-2"x6"x10'-11"
LONG WEB 2"x4"x4'-9"
SHORT WEB 2x4'x2'-9"
LOWER CHORD 2"x4"x9'-0'
10'-11"
3'-0"
4'-9"
2"x4"x4'-0"
TIE
12
4
3'-7" DRESSED
3'-8" FULL SIZE
APPROX.
6'-0"
6" 6"
MIN
18'-4" SPAN (OUTSIDE OF PLATE)
MIN
6 6
8-2 1/2" SR
4-1/2"x6" BOLTS
6-2 1/2" SR
3-1/2"x6" BOLTS
LOWER CHORD
SPACER BLOCK
2"x4"x3'-0"
4/12 ROOF TRUSS (S.R = SPLIT RING CONNECTOR)
SCALE: 1/2"=1'-0"
LONG WEB
SHORT WEB
UPPER CHORD
1"x6" SWAY BRACES FROM TOP OR BOTTOM OF ONE TRUSS TO BOTTOM OR TOP OF ADJACENT TRUSS
TIE
SPACER BLOCK
LOWER CHORD
ISOMETRIC OF ROOF TRUSS
ROOF GIRT NAILED TO 2"x4" SUPPORT
2"x4"
METAL FASTENER
DETAIL OF ALTERNATE METHODS OF FASTENING ROOF GIRT TO TRUSS
1"x4"
1"x4" STOP
2"x6"
DOOR
SECTION F-F
SCALE: 1"=1'-0"
*P.P.T. - PRESSURE PRESERVATIVE TREATED

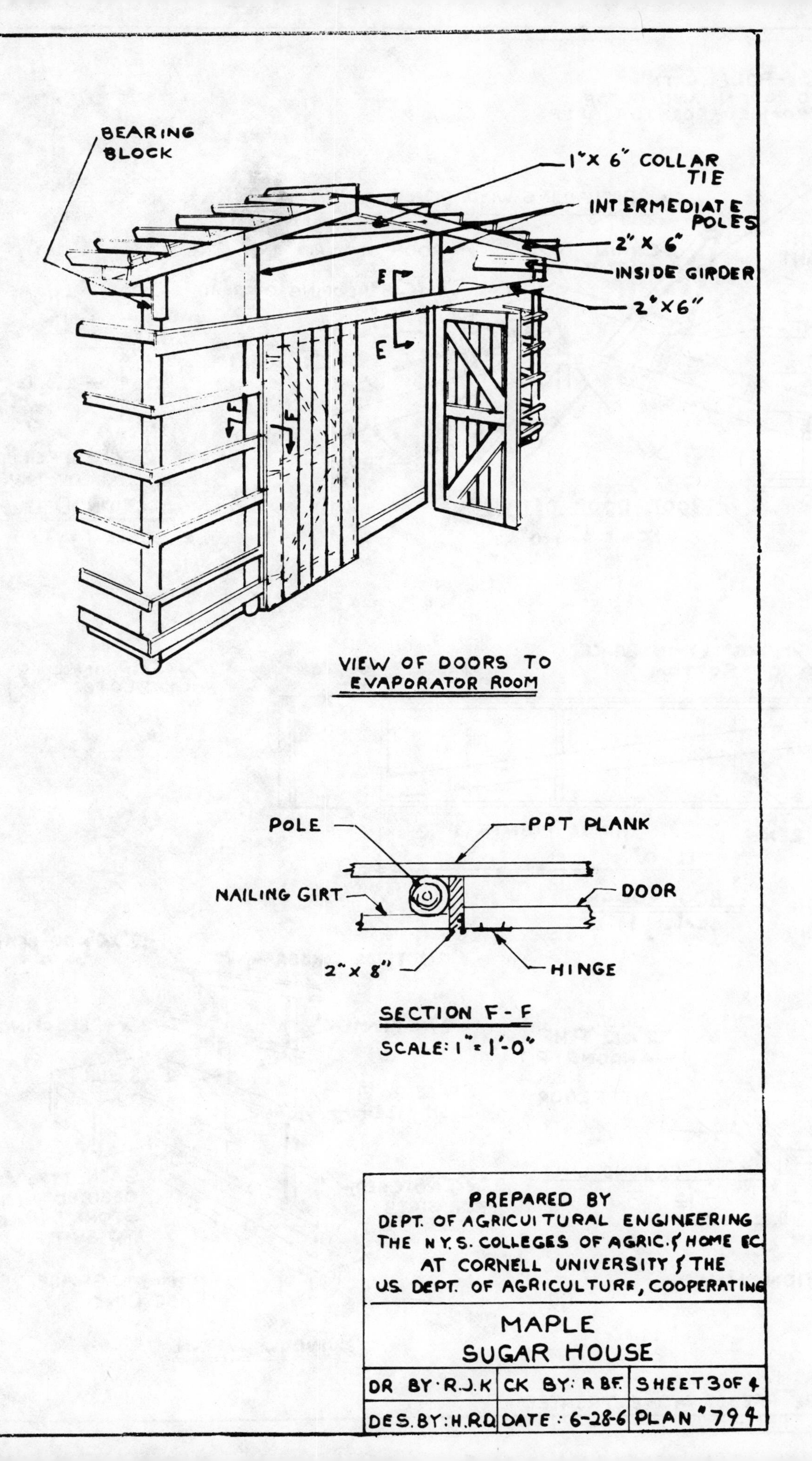
BEARING BLOCK
1"x 6" COLLAR TIE
INTERMEDIATE POLES
2" x 6"
INSIDE GIRDER
2"x6"
E
E
F
F
VIEW OF DOORS TO EVAPORATOR ROOM
POLE
PPT PLANK
NAILING GIRT
DOOR
2" x 8"
HINGE
SECTION F - F
SCALE: 1" = 1'-0"
PREPARED BY
DEPT. OF AGRICULTURAL ENGINEERING
THE N.Y.S. COLLEGES OF AGRIC. & HOME EC.
AT CORNELL UNIVERSITY & THE
U.S. DEPT. OF AGRICULTURE, COOPERATING
MAPLE
SUGAR HOUSE
DR BY: R.J.K
CK BY: R.B.F.
SHEET 3 OF 4
DES. BY: H.R.D.
DATE: 6-28-6
PLAN #794

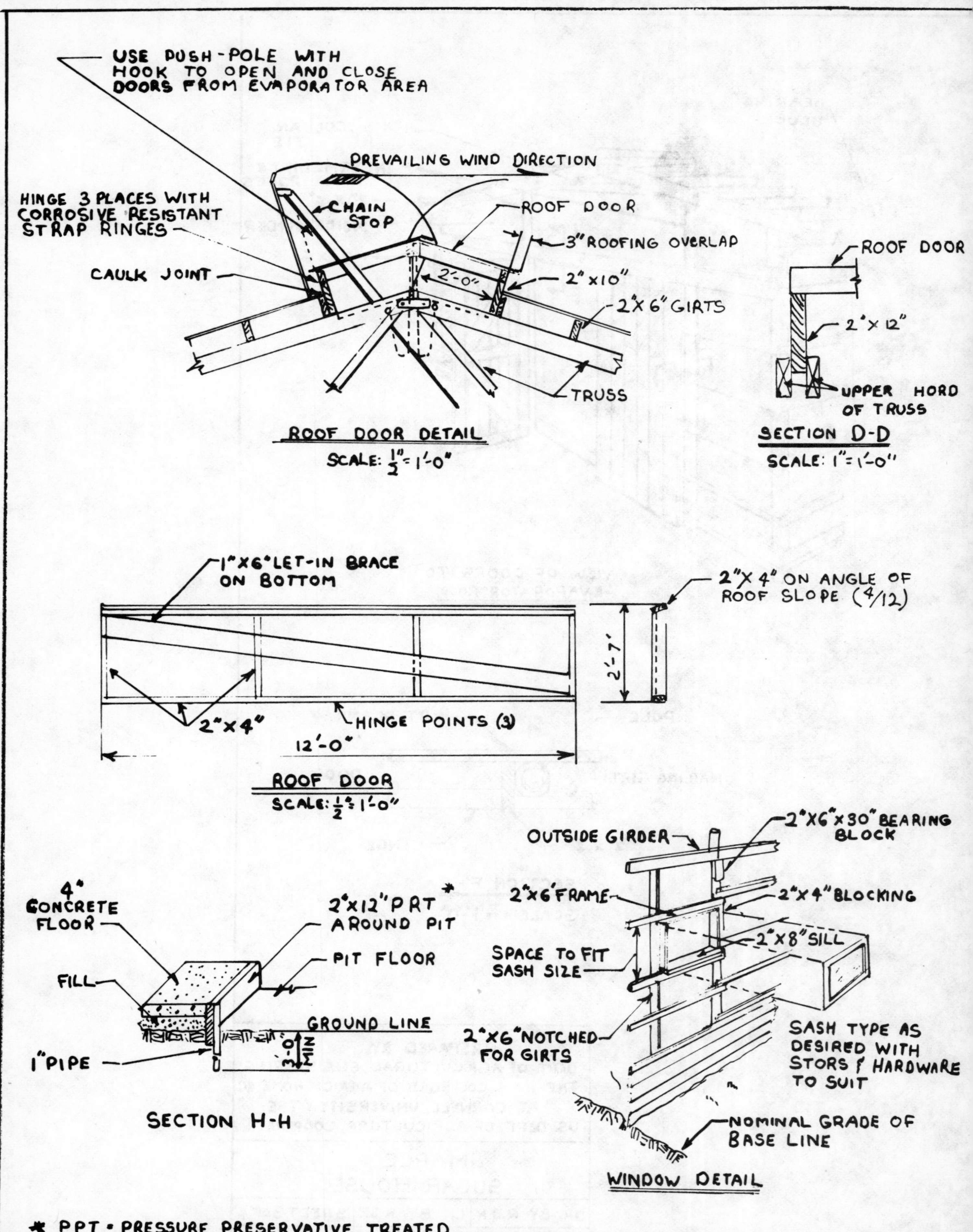

USE PUSH-POLE WITH HOOK TO OPEN AND CLOSE DOORS FROM EVAPORATOR AREA
PREVAILING WIND DIRECTION
HINGE 3 PLACES WITH CORROSIVE RESISTANT STRAP RINGES
CHAIN STOP
ROOF DOOR
3" ROOFING OVERLAP
CAULK JOINT
2'-0"
2" x 10"
2" x 6" GIRTS
TRUSS
ROOF DOOR DETAIL
SCALE: 1/2" = 1'-0"
ROOF DOOR
2" x 12"
UPPER HORD OF TRUSS
SECTION D-D
SCALE: 1" = 1'-0"
1" x 6" LET-IN BRACE ON BOTTOM
2" x 4" ON ANGLE OF ROOF SLOPE (4/12)
2'-7'
2" x 4"
HINGE POINTS (3)
12'-0"
ROOF DOOR
SCALE: 1/2" = 1'-0"
OUTSIDE GIRDER
2" x 6" x 30" BEARING BLOCK
2" x 6' FRAME
2" x 4" BLOCKING
SPACE TO FIT SASH SIZE
2" x 8" SILL
2" x 6" NOTCHED FOR GIRTS
SASH TYPE AS DESIRED WITH STORS & HARDWARE TO SUIT
NOMINAL GRADE OF BASE LINE
WINDOW DETAIL
4" CONCRETE FLOOR
2" x 12" PRT * AROUND PIT
PIT FLOOR
FILL
GROUND LINE
1" PIPE
3'-0" MIN.
SECTION H-H
* PPT - PRESSURE PRESERVATIVE TREATED

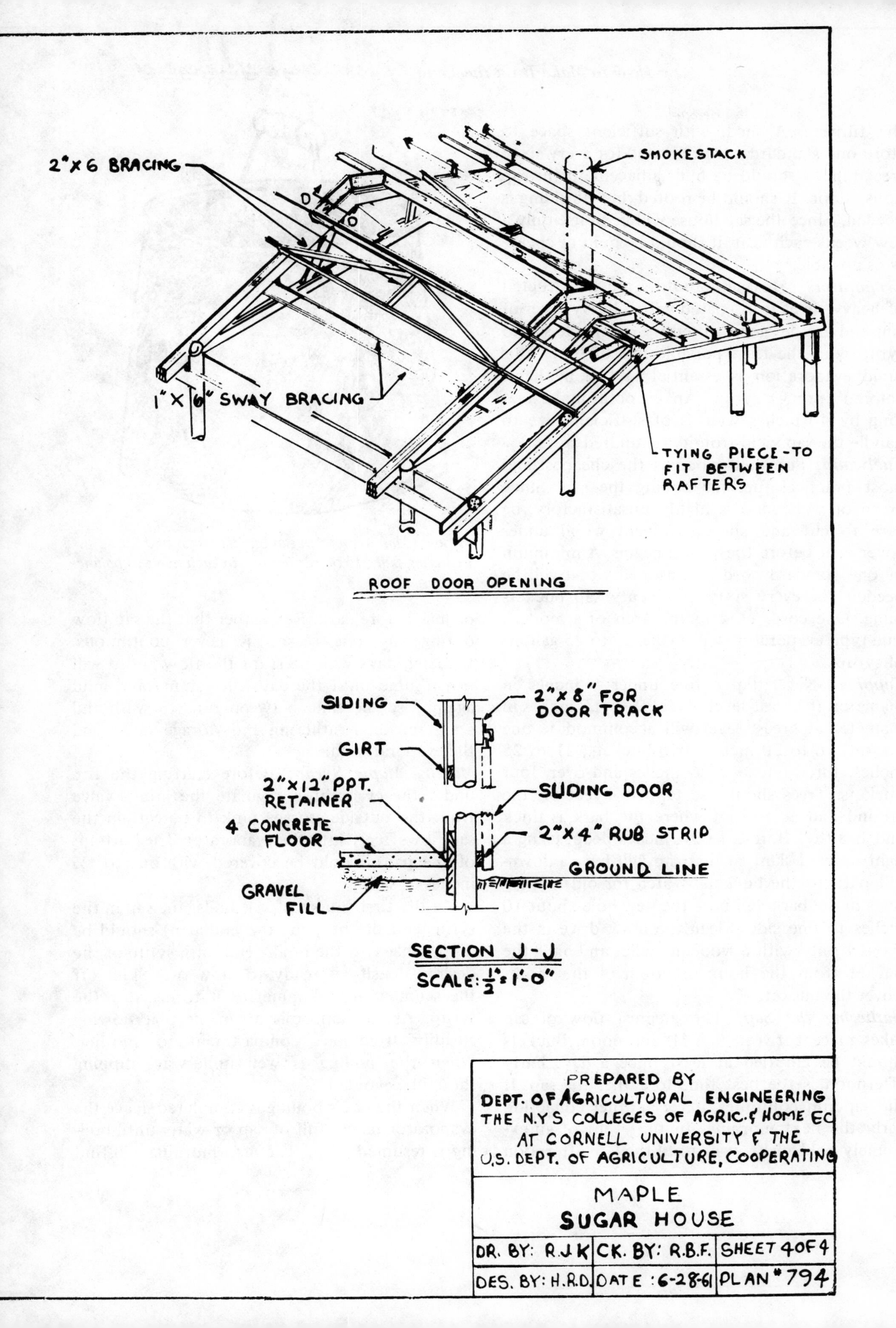
2"X 6 BRACING
SMOKESTACK
1" X 6" SWAY BRACING
TYING PIECE-TO
FIT BETWEEN
RAFTERS
ROOF DOOR OPENING
SIDING
2"X 8" FOR
DOOR TRACK
GIRT
2"X12" PPT.
RETAINER
SLIDING DOOR
4 CONCRETE
FLOOR
2"X 4" RUB STRIP
GROUND LINE
GRAVEL
FILL
SECTION J-J
SCALE: 1/2" = 1'-0"
PREPARED BY
DEPT. OF AGRICULTURAL ENGINEERING
THE N.Y.S COLLEGES OF AGRIC. & HOME EC
AT CORNELL UNIVERSITY & THE
U.S. DEPT. OF AGRICULTURE, COOPERATING
MAPLE
SUGAR HOUSE
DR. BY: R.J.K
CK. BY: R.B.F.
SHEET 4 OF 4
DES. BY: H.R.D.
DATE: 6-28-61
PLAN # 794

the furnace. A shed, with sufficient space to store one standard cord of wood for every sixty trees tapped should be built adjacent to the sap house door. It should be roofed, but no siding is needed. Since the sap house will be in use only a few weeks each year, it should be built as cheaply as possible.

Evaporators: Most evaporators are constructed of heavy sheet iron coated with a heavy layer of tin, and are designed to reduce the maple sap to syrup with the least possible loss of time. This rapid evaporation is essential in making light-colored (prime) syrup. An evaporator 12 feet long by 40 inches wide is of sufficient size to handle the sap yield from two hundred trees.

Fuelwood: Seasoned wood is the cheapest and most practical fuel for boiling the sap. Since green or wet wood is highly unsatisfactory, be sure to cut and stack sufficient wood under cover well before the season begins. A minimum of one standard cord of seasoned wood will be needed for every sixty or seventy sap buckets hung. One cord, used in the arch of a modern flue-type evaporator, will make 20 to 25 gallons of syrup.

Tapping: Never tap a tree under 8 inches in diameter at breast level. Trees 9 to 15 inches in diameter at breast level will accommodate one bucket; 16 to 20 inches, two buckets; 21 to 25 inches, three buckets; 26 inches and over, four buckets. Trees should be tapped 4 feet off the ground and at a point where the bark is thick and healthy. Bore 2 to 2½ inches deep, using a light upward slant so that sap will have a downhill path to the buckets. Watch for old tapping scars in the bark and bore the new hole about 10 inches to one side. Clean the bore, drive in the spout firmly with a wooden mallet and hang the bucket from the hook or notch of the spout. Cover the bucket.

Gathering the Sap: The greatest flow of sap takes place between 9 A.M. and noon. Buckets should be emptied at least once a day. Early afternoon is the best time to gather the sap. If the sap run continues, make another collection early the next morning. In the event of an extremely cold night, wait until the late afternoon

(Above) *Maple sap from sap buckets is poured into gathering pail, to be carried one in each hand to balance load.*

or just before dark. Remember that the sap flow during any one season is never continuous. Freezing days will interrupt the flow and it will not resume until the days turn sunny and mild again. Use the time between runs to wash and scald buckets, gathering and storage tanks and the evaporator pan.

Making Maple Syrup: Before starting the fire under the evaporator, regulate the intake valve from the outside storage tank to be certain the sap flows freely into the evaporator. The bottoms of the pans should be covered with one to 1½ inches of sap.

As the first intake of sap boils, the sap in the syrup end of the pan (the end pan) should be dipped back to the intake end until syrup of the desired density is ready to draw off. Skim off the scum with a skimmer as it forms atop the syrup. As the sap boils down, keep it flowing steadily from one compartment to another. When the boiling is well underway, dipping should be avoided.

When the day's boiling is completed, leave the evaporator partly full of sap or water until boiling is resumed again. The next morning will find the bottom of the pan covered with a deposit which should be loosened and the sap stirred before building the new fire. Each time the evaporator is started, care should be used to keep the pans from boiling dry and to insure a constant flow from sap intake to syrup outlet. It is important to clean the evaporator every two to three days.

An essential item for making maple syrup is a syrup thermometer. Since the thermometer is set to read at sea level and very few sap houses are at that level, it will be necessary to correct it before using. A simple method is to bring a bucket of snow water to boiling, put the thermometer in the water, read, and add seven degrees to the reading. This will give you the proper boiling point for the syrup. Using a Baume Hydrometer also comes in handy for checking syrup density. The Baume reading for finished syrup is 36° at 60° F. Syrup should weigh between 11 and 11¼ pounds to the gallon for table use.

Scoop for siphoning off ready syrup at production end of maple syrup evaporator for testing with Baume Hydrometer for quality.

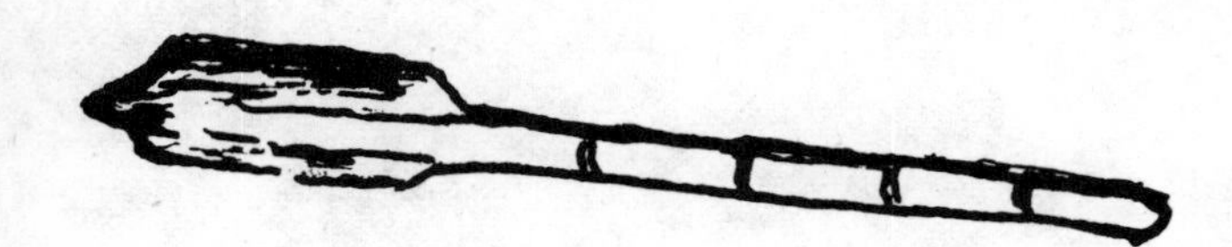

Baume Hydrometer

During the boiling process it is necessary to cleanse the syrup of impurities. The white of an egg, cream, milk or cold sap may be added to the boiling sap. This action will bring the impurities to the surface where they can be skimmed off and discarded.

Fine-quality maple syrup is being drawn from evaporator for canning or bottling.

Filtering the Syrup: When the hot syrup is drawn from the evaporator it should be filtered through heavy felt filters of one-gallon capacity. Filters remove sediments from the syrup. Cotton flannel may be substituted for the felt. Filters should be kept clean by washing frequently. Suspend the filters in the mouth of the settling tank and run the hot syrup through it directly from the evaporator.

Canning and Marketing: Syrup from the settling tank can be poured hot or cold into cans, bottles or jars. The highest prices for maple syrup and maple syrup products may be achieved by bottling in one-quart to ½-pint containers for the retail trade. Glass containers are preferable to metal cans because the consumer likes to see what he is buying. Design your own labels and sell your maple products directly to the consumer. There is a ready market in roadside sales, a mail-order business or retail route, especially when combined with the sale of other farm products. Another ready market is the "Country Stores," springing up in just about every city in the nation.

Making Maple Sugar: Maple sugar is usually prepared in the sap house in a special "sugaring-off pan." It can be made into a large variety of hard and soft sugars and packed attractively by using a variety of fancy molds.

Maple sugar requires an extremely hot fire, and great care must be used to prevent it from burning. Hot sugar has a tendency to foam up and boil over, so the maker must be able to give his full time to sugaring off. An antifoaming agent can be purchased and is useful at this stage.

To make soft or tub sugar, boil the syrup until the thermometer registers from 26 to 28 degrees above the temperature of boiling water (from 238° F. to 240° F. at sea level). If the boiling is continued until the thermometer registers from 30 to 33 degrees above the temperature of boiling water, the resulting product will be hard or cake sugar. Often enough the top will fall in after the hot sugar has been poured into cans or pails. To overcome this, fill containers no more than three-quarters full the first time, then top them off at the next sugaring off with the same grade of sugar.

Maple Syrup for Home Consumption: About the only equipment needed to make maple syrup for home consumption is a 3/8-inch bit and carpenter's brace; syrup thermometer, spouts, sap buckets or ordinary pails, a kettle or cauldron, and enough cans, bottles or jars to put the syrup up in. If you don't have buckets or pails, use 2-pound coffee cans, with a hole punched into one side close to the rim to accommodate the hook of the spout.

When the weather is ripe, tap a few trees, using the same methods as already stated in this chapter. Collect the sap and pour it immediately into a large kettle or cauldron set over a very hot fire. Boil until it registers the proper degree on the thermometer. Strain into a large container, using a double fold of cheesecloth. Clarify by adding an egg white or one half cup of milk. When sediment rises to the surface, skim it off. Cool, pour into sterilized cans or jars. Seal containers.

8

Maple Products

The production of maple syrup and its many by-products presents a golden opportunity to the full-time and part-time farmer to harvest a cash crop at a time of the year when very little other work can be done on the farm. The season is short, the work is pleasant, and the rewards are more than worth the effort involved.

Most large producers still sell their maple syrup to wholesale canners for approximately $6 per gallon. This same gallon of syrup, converted to maple products, attractively packaged and marketed on a farm-to-consumer or farm-to-retailer basis, will bring the producer three times or more the wholesale price.

Making different maple syrup products presents no special difficulty, nor does it call for expensive or unusual equipment. It does require the same type of sanitary conditions and care that are expected of any producer of food products for sale to the consumer.

A separate room in the house, a closed-off porch or a partitioned-off area in the sap house will do nicely as a working area for making maple confections.

Unlike maple syrup, which is boiled over roaring wood fires, maple confections require the more even, easily controlled heat that only gas, bottled or natural, can furnish. Other equipment needed includes kettles, mixers, pans, wooden ladles and paddles, measuring cups, a household scale and a candy thermometer, all of which may be purchased in your local hardware or houseware store.

Gallon can for maple syrup. For farm-to-road traffic sales maple syrup should be put up into glass jars or jugs.

Maple Cream

Maple cream is a soft, fondant-type confection that spreads as easily as butter, and has become so popular in recent years that many producers of maple syrup convert their entire crop to the confection.

Maple cream is made up of millions of microscopic sugar crystals interspaced with a thin coating of saturated syrup (mother liquid). The crystals are impalpable to the tongue and give the cream its smooth texture. The first step in making maple cream is to produce a supersaturated sugar solution, using Fancy or Grade A maple syrup.

Cooking and Cooling: Heat the syrup to a temperature of 22° to 24° F. above the boiling point of water, immersing the thermometer in a pot of boiling water to calculate the correct reading. As soon as the boiling syrup reaches the correct temperature, remove it immediately from the heat and cool quickly, to prevent crystallization.

For quick cooling, use large, flat-bottom pans, and pour in the syrup to a depth of not more than one to 3 inches. The pans should be placed in a trough through which cold water (35° to 45° F.) is flowing, or in a larger pan of water with enough ice added to cool the syrup to a temperature of at least 70° F., and preferably 50° F. It will be cooled sufficiently when the surface is firm to the touch.

Creaming: The chilled, thickened syrup may be creamed by hand or by a mechanical cream beater, of which a number of inexpensive models are on the market. When creaming by hand, pour the cooked syrup into a large flat pan such as a cookie tin. Using a hardwood paddle with a sharp edge 2 or 3 inches wide, hold the pan firmly in one hand and scrape the thick syrup, first to one side of the pan and then to the other. Mixing should be continuous. If stopped, some of the crystals may grow and make the cream gritty.

While being creamed, the chilled syrup first tends to go into a fluid stage and then begins to stiffen and display a tendency to set. At about this time the product loses its shiny appearance. One way to hasten the creaming process is to "seed" the chilled syrup just before beating. The "seed" in question is a small amount of previously made cream. Add one teaspoon of this "seed" to one gallon of cooked syrup, and the creaming process, which normally takes one to 2 hours, depending on the size of the batch, can be cut in half.

Packaging and Storing: The best type of containers for maple cream are widemouthed glass jars of ½-pint capacity. When filling, care should be taken to keep air bubbles from forming because they give the customer the impression that the package is short in weight.

Freshly made cream may be packaged immediately, set aside covered for one day to "age," then remelted and packaged while it is still warm and fluid. The cream is best stored at low temperatures, under refrigeration, but a springhouse, with a constant temperature of approximately 60° F. or below, will do just as well. Packaged in glass or other moisture-proof containers, the cream can be stored for long periods without danger of the saturated syrup separating from the cream.

Fondant

Fondant is a confection very similar to maple cream because of its fine crystalline character. The method of making the confection is the same as in making maple cream except that the syrup is heated to a higher boiling point (27° F. above the boiling point of water instead of 22° to 24° F.). The thickened syrup should be cooled to 50° F. and then stirred, as for creaming. Since there is less syrup left in the fondant once creaming has been completed, the confection will set up to a soft solid at room temperature. Drop small amounts on waxed paper, a cookie tin or a marble slab, or transfer from the mixing pan directly into molds for packaging.

Soft Sugar Candies

Next to maple cream, maple soft sugar candies have found the readiest market among consumers over the last few years. Like maple cream and fondant, 8 pounds of soft sugar candies can be made from one gallon of maple syrup.

Maple candies are much stiffer in texture than maple cream because they contain little or no free syrup.

Cooking, Cooling and Stirring: Syrup for maple candies should be cooked to a temperature reading of 32° F. above the boiling point of water. Cool the pans of cooked syrup slowly to 155° F., as tested with a thermometer. The thick syrup should then be stirred, either by hand with a large wooden spoon or with a mechanical mixer.

When the sugar becomes soft and plastic, pour into rubber molds of various shapes (the most popular of the molds used in today's market are leaf-shaped). They should be filled with a spatula or wide-blade putty knife. The molds, along with everything else needed for making maple syrup and its by-products, may be purchased at reasonable prices from your local supplier of maple production equipment.

Before using the molds, be sure to wash them in a strong solution of alkali soap, rinse them under running water and dry thoroughly. When dry, coat with glycerine applied with a clean pastry or paintbrush. Remove any excess glycerine by blotting thoroughly with a soft clean cloth.

Maple Fluff

Maple fluff is made from a lower quality, darker grade of syrup than is used for the other maple products described. It is easier to make than maple cream, has the same spreading consistency, is packaged in the same manner as maple cream and is just as popular with consumers as its lighter-colored sister.

To make maple fluff, heat the syrup to a temperature of 17° F. above that of boiling water. Allow it to cool, with only occasional stirring, until the thermometer registers between 185° and 175° F. Add 1/3 cup of purified monoglyceride (purchase at drugstore or from local supplier of maple syrup equipment) per gallon of syrup a little at a time and stir continuously until it is thoroughly dissolved in the hot syrup. Cool to a temperature of between 160° and 150° F. and whip the mixture quickly, using a high-speed household beater. Fluffing should occur within 2 minutes. Package as in maple cream.

Fluff From Maple Syrup and Maple Sugar

To one cup of maple syrup add ½ cup of maple sugar and heat the mixture until the sugar is completely dissolved. *Do not boil.* Cool to 175° F., stirring only occasionally. Add slowly, and stir until dissolved, one teaspoon of monoglyceride for each cup of syrup. Cool to 150° F., and whip the mixture with a high-speed household beater until it fluffs. Fluffing should occur within 2 minutes. Package as in maple cream and maple fluff.

Hard Maple Sugar

Because it is not too easy to eat, hard maple sugar is not usually classified as a confection, but producers do find there is a demand for hard sugar because it offers a convenient form for the storage of maple syrup. The hard sugar cake can be broken up and melted in water and the solution boiled to bring it up to syrup density. This syrup is called maple sugar syrup to distinguish it from syrup made directly from sap.

To make hard sugar, heat the desired amount of maple syrup to 40° to 45° F. above the boiling point of water. When the syrup reaches the desired temperature, remove it from the heat and stir until the syrup begins to crystallize and stiffen. Pour into molds or wooden tubs for storage.

Maple Syrup on Snow

This is a great favorite with children and grown-ups alike during syrup-making time in the maple orchards and sap houses. After the syrup is drawn from the evaporator, it is heated to approximately 40° F. above the boiling point of water, then poured immediately, without stirring, onto pans of snow or chipped ice. The mixture is cooled rapidly so that the supersaturated syrup has no chance to crystallize. The result is a thin, glassy, taffy-like sheet of delicious maple candy.

Maple Vinegar

Maple syrup skimmings (from the evaporator), dark or slightly scorched syrup, and even maple sap or syrup which has just begun to ferment will make excellent vinegar.

30 gallons Maple syrup diluted with water (soft) to weigh 9 pounds per gallon
2 ounces Ammonium sulphate
2 ounces Sodium sulphate

Dissolve the chemicals in the water used for diluting before adding it to the maple syrup. The water should be warm and the mixture added to hot, but not boiling, syrup. Diluted syrup, prepared as above, should be inoculated with approximately ½ cup of old vinegar, carefully strained. Keep liquid at about 80° F. until vinegar of desired taste has been achieved.

Note: "Old vinegar" can be procured from any neighboring maple syrup operator.

9

Beekeeping

Before we start talking about how to raise bees, let us take a look at beekeeping as an industry and at the busy little fellows themselves. Who knows? Maybe we'll both learn something we did not know before.

In our country beekeeping has been a traditional craft since pioneer times—when almost every family had "bee-gums" to provide "sweetening." Their importance in producing honey, so rich in mineral content and quick energy, has not diminished with the passing years.

Beekeeping as an Industry

Starting with the state of New York, where about nine thousand persons keep a total of around 160,000 colonies of honeybees, the annual production is about 9,300,000 pounds of honey and 155,000 pounds of beeswax. These figures are higher than the national average of state production, but not as high as those of leading honey states in the Midwest. New York ranks first in the East in honey and is usually among the first seven in the country in the number of colonies plus production.

A completely equipped colony of bees costs from $30 to $45, which means that more than $5 million is invested in bees and equipment, exclusive of honey houses and honey packing plants, in New York alone.

The Honey Bee as a Pollinator

The pollination of agricultural crops is the most important contribution of honeybees to our national economy. Though it is impossible to estimate the value of honeybees for pollination, it is many times the total value of both the honey and beeswax which they produce. Without cross-pollination many crops would not set seed or produce fruit. Many insects other than the honeybee may carry pollen from one plant to another, but in areas where agriculture has been intensified the number of these other insects is inadequate to large-scale pollination.

Several conditions have contributed to a decline of the native pollinating insects in certain areas. In recent years there has been a trend toward intensive and specialized agriculture. General or diversified farming, as I have mentioned, is neither popular nor really profitable. But land used to grow only one crop does not provide nectar and pollen for wild pollinating insects over a long period of time. The elimination of hedgerows when fields are made larger reduces the nesting locations for pollinating insects. Pesticides also have their effect.

It is estimated that over fifteen thousand colonies of honeybees are used to pollinate apples and other fruits in New York State alone. Bees from New York are moved to Vermont, Massachusetts, Connecticut and Maine for the same purpose. While fruit pollination is the primary use for which beekeepers are paid, about one thousand colonies are rented for the pollination of other agricultural crops, including cucumbers and cantaloupes. Beekeepers tend to think in terms of *crops* from which they receive compensation, but conservationists are aware of the pollination of wild fruit, nut and seed crops which also benefit from colonies of honeybees.

Who Keeps Bees?

Bees are kept by persons in all walks of life. For some, beekeeping is an interesting hobby; for those who operate several thousand colonies, it is an important source of income. Many women keep a few colonies, but seldom more because of the heavy lifting.

A Skilled Occupation

Beginners in beekeeping are often motivated by a desire to make a living from honey production. So many things are involved that it is impossible to state the number of colonies needed to reach this goal. If bees are to be given proper attention, five hundred colonies are probably the maximum a skilled beekeeper can manage without paid help.

The complete equipment necessary to operate such an outfit efficiently may require an investment of $15,000 to $20,000. No beginner should consider such an outlay until he or she has had several years experience with bees and is sure of the necessary ability to produce and sell enough honey to make the venture a financial success. A good plan is to increase the number of colonies each year while continuing a full- or part-time job.

Acquire a Knowledge of Beekeeping First

Success with even a few colonies requires a thorough knowledge of the life and behavior of bees. A good way to obtain this knowledge and at the same time learn whether beekeeping is an occupation *you* would like is to work with a skilled beekeeper. While getting practical experience, spare time may be devoted to reading the best books and bulletins on the subject. Few persons make a success of beekeeping without really *practical* experience.

National Organizations

The American Beekeeping Federation represents the *entire* beekeeping industry. The place of the annual meeting is changed from year to year to provide beekeepers in different sections of the country an opportunity to attend. Announcements concerning this organization are made in many bee journals, a few of which you should subscribe to.

The American Honey Institute is an organization sponsored by the beekeeping associations, bee supply houses, honey packing plants and individual beekeepers. Its purpose is to study new and better uses for honey in the home, to prepare reliable literature and to give publicity to honey. Recipe booklets and pamphlets on the use of honey in the home may be obtained from the home office of this organization at 333 North Michigan Avenue, Chicago, Illinois 60601.

Start Your Beekeeping When . . .

The best time is in the spring about when the dandelions and fruit trees begin to bloom. Many beginners start by buying packages which consist of 2 or 3 pounds of bees and a mated queen shipped in a temporary wire cage without combs from the southern states.

It is advisable to install package bees on full sheets of "comb foundation" instead of "drawn combs," and to feed them sugar syrup instead of honey. These precautions are taken to guard against the possibility of spreading American foulbrood. The same precautions should be taken in hiving stray swarms of bees. Established colonies purchased from a neighboring beekeeper should be well supplied with honey and should be in standing hives in good condition and accompanied by a certificate of disease inspection from your State Apiary Inspector, Department of Agriculture and Markets, capital of your state.

Buying Bees and Queens

Package bees and queens may be purchased from southern producers, most of whom advertise in the beekeeping journals. Established colonies and equipment for sale are also advertised in these journals.

Italian bees are recommended for beginners because they are the most commonly used by beekeepers in this country and are raised by practically all of the queen and package-bee producers. A few breeders raise Caucasian and Carniolan bees, but most consider Italian bees the best.

Literature on Beekeeping

There are hundreds of books on beekeeping in this and other countries. They may be obtained from publishers, libraries, bookstores and bee supply dealers.

Bulletins and circulars on specific phases of beekeeping are available from a variety of sources; the states, Canadian provinces and federal governments publish separately and independently on a variety of special areas.

Beekeeping trade and society journals are published in this country and abroad. Their interests vary widely; some cater to hobby beekeepers and others to commercial interests.

One valuable source of information is *The Honey Market News*, issued monthly in Washington, which deals with conditions of colonies and honey plants, prices of honey and beeswax in various markets and other information of interest to beekeepers. It may be obtained without cost by writing to the Consumer and Marketing Service, Fruit and Vegetable Division, Washington, D.C. 20250.

Equipment Needed

The modern beehive is made up of several parts. The "bottom board" supports a "hive body" which is topped with an "inner" and an "outer cover." The hive body contains ten "frames," each of which encloses a full sheet of foundation.

Foundation is a thin sheet of beeswax that has been embossed with the octagonal pattern common to all honeycomb. It is used because the resulting comb is straight and thus easy to remove.

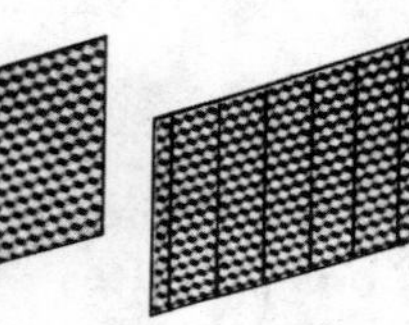

Supers in 3 depths. Shallow for chunk or extracted honey; medium for extracted; deep for extracted or brood chamber. 20x16¼ in. wide. Unpainted pine, dove-tailed corners. Nails and assembly instructions.

Frames. 17⅝ inches long, divided bottom bar; cut-out cleat for easy foundation insertion. Top bar 19x¾ inches wide. Unpainted pine. Nails and assembly instructions included.

100% pure Beeswax Comb Foundation. Pressed cell pattern encourages building of perfect combs, tends to restrict drones.
Use for chunk honey . . tender texture makes it tempting to the taste.

Wire-reinforced Comb Foundations. Crimped steel wires embedded in 100% pure beeswax. Pressed cell pattern.
Strong, reusable, resists sagging. Use with divided bottom frames.

In addition to the complete hive and a package of bees, the beekeeper should have at least three "supers" in which to produce honey. Each super contains ten frames and ten sheets of foundation.

A beginning beekeeper will also need a veil to protect the face and a pair of specially constructed bee gloves.

Two other items complete the equipment list: a "smoker" and a "hive tool." The smoker is a small canister with an attached bellows. When it is filled with material such as burlap and the burlap is ignited, the beekeeper can direct smoke over the frames and subdue the bees. The hive tool is a flat piece of steel about 10 inches long, curved on one end, and is used to pry open the hive, raise frames and accomplish other tasks.

Homemade Equipment

Many commercial beekeepers believe that it pays to buy accurate factory-made equipment. The frames fit better in the hives and require less effort for removal and replacement. On the other hand, beginners often want to make their own equipment. If hives are to be made at home, the best plan is to buy a complete hive for a model. Exact dimensions must be adhered to or the bees will build comb and deposit propolis (a resinous, adhesive substance produced by bees to serve as a cementing material—bee glue) where it is not desired.

Where to Place Apiaries and How Far Apart

Colonies should be placed in a sunny area sheltered from prevailing winds. The area should also be well drained. The bees should be kept far enough from the road to prevent them from becoming a nuisance to passersby. A hedge may help in this regard. When possible, have entrances face south or southeast.

It is an unwritten law among beekeepers that no sizable apiary should be installed within 2 miles of another one. If this practice is violated, *both* apiaries usually produce a smaller total crop of honey.

(Right) *Bee colony*

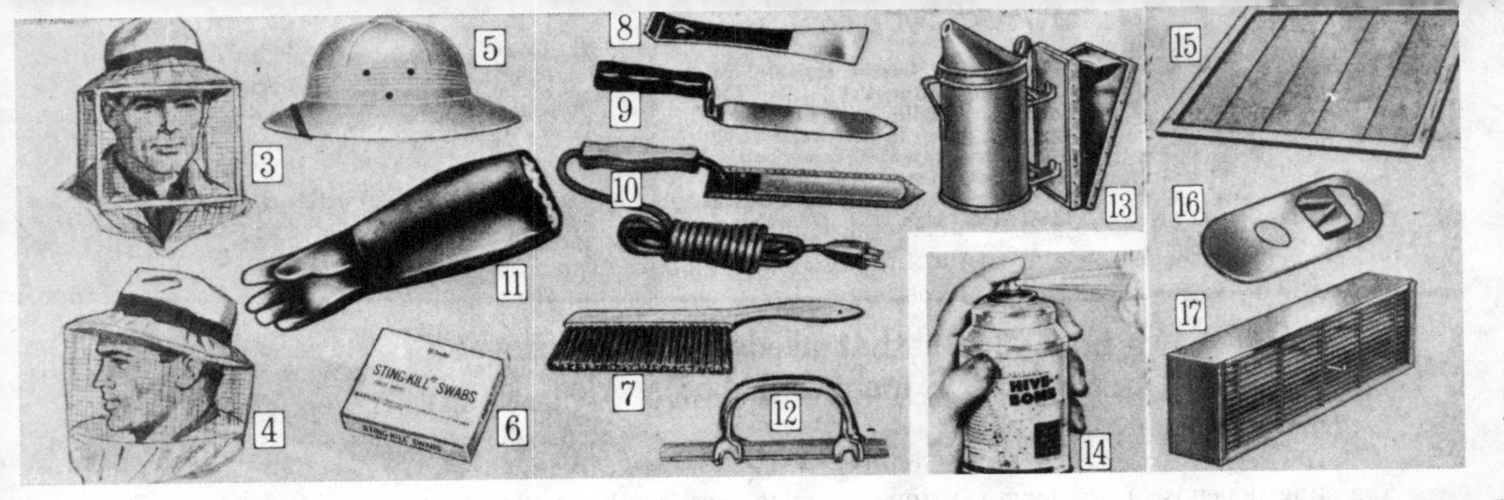

and 4) **Steel mesh Bee Veils.** sticized top . . drawstring bottom.

Folding Veil. Stays smooth.
32 KF 55212—Shpg. wt. 12 oz.$2.95

Canvas-bound Veil. Easy to fit.
32 KF 55211—Shpg. wt. 8 oz. . $1.89

Fiber Helmet. Adjustable head-band. Stiff brim. Ventilated.
KF 55213—Shpg. wt. 2 lbs. . . . $1.95

Sting-Kill®. 10 antiseptic swabs.
32 KF 55271—Wt. 2 oz. . . Box $1.69

Black horsehair Brush. 14 in.
32 KF 55225—Shpg. wt. 6 oz. . . 95c

Steel Hive Tool for removing propolis. Nail slot. 8-in. blade.
KF 55224—Shpg. wt. 4 oz. . . . $1.45

9 **Uncapping Knife.** 10-inch polished steel blade slices clean.
32 KF 55232—Shpg. wt. 1 lb. . . . $3.95

10 **Electric Uncapping Knife.** Thermostat holds proper heat to keep honey from sticking to 10-inch double steel blade. 5-foot, 3-wire cord. 660 watts. 110–120-volts, 60-cycle AC. *UL listed.*
32 KF 55231—Shpg. wt. 3 lbs. . . $21.50

11 **Treated Cotton Gloves with Plastic-coated hand**—prevents bees from stinging hands. Elbow length. Elasticized top. Large.
32 KF 55215—Wt. 1 lb. . . . . Pair $3.75

Flannel-lined Cotton Gloves, resistant to stings. Similar to above. Large.
32 KF 55214—Wt. 1 lb. . . . . Pair $1.75

12 **Aluminum Frame Grip.** One-hand lift, 8 inches long.
32 KF 55223—Shpg. wt. 1 lb. . . $2.30

13 **Bee Smoker.** Tinned steel, bellows. Hinged top. 4x7 inches.
32 KF 55222—Shpg. wt. 2 lbs. . $3.95

14 **Push-button Hive Bomb.** Hardwood smoke concentrate helps quiet bees. Use from 25 to 30 times.
32 KF 55221—Shpg. wt. 1 lb. . . $1.65

15 **All-wire Excluder.** Keeps drones and queen out of supers. Wood frame. For standard hives.
32 KF 55243—Shpg. wt. 2 lbs. . $2.10

16 **Bee Escape.** Prevents bees' return to super.
32 KF 55242—Shpg. wt. 2 oz. . . . 59c

17 **Queen-Drone Trap.** Wire grille lets out workers only. Place in hive entrance at swarming time. Fits any standard size hive (20x 16¾x9⅝ inches high).
32 KF 55241—Wt. 1 lb. 2 oz. . . $2.10

Bee Stings

Bee stings are annoying to most experienced beekeepers as well as beginners, but the degree of sensitivity varies considerably. Experienced beekeepers still suffer the same pain from the initial prick of a sting, but they quickly build up an immunity to the bee venom, which reduces or eliminates aftereffects.

The degree of pain and swelling resulting from a sting depends on the length of time the sting remains in the skin and the amount of poison that enters the flesh. It is therefore important to rub or scrape out the sting immediately; never pull it out, for the fingers press the poison sacs attached to the base of the sting and force the poison through the barbs into the flesh.

Bees and other stinging insects seem to be disturbed by the odor of sweat and most pomades and perfumes, and are impelled to sting. They are not likely to sting light-colored, smooth-textured clothing. Beekeepers should therefore prepare for work in the apiary by washing in nonfragrant soap and dressing in clean, white coveralls or khaki clothing that covers their bodies as much as possible.

Bee Poisoning

The poisoning of bees is not too common, but it does happen, especially in areas where competition for pollinating bees is strong. In case of suspected poisoning of adult bees, contact your county Cooperative Extension agent and try to determine as quickly as possible the type of poison that was used. At least two hundred sick or dead bees for each type of poison suspected should be sent for examination and analysis to The Apiculture Research Branch, Entomology Research Division, Beltsville, Maryland 20705. Mail all samples in a wooden or heavy cardboard box; never use tin, glass or waxed paper, because the samples usually mold.

Other Dangers

The wax moth (*Galleria melonella*), often referred to as the bee moth, wax worm and web worm, is found throughout the country and causes severe damage to combs. It lays eggs in the crevices of the hives. On hatching, the minute, highly mobile larvae make their way to the combs. The larvae grow rapidly and construct strong tunnels of silk as they burrow through and among the combs, feeding on the pollen, cocoons and honey. On reaching maturity the larvae spin tough silken cocoons around themselves, emerging later as adult moths.

During warm weather, weak colonies as well as combs in storage are subject to wax-moth attack. Destruction is rapid and so complete that the combs are reduced to a mass of webs and debris in a few weeks.

Stored combs should be examined every two weeks. When the first signs of wax moth appear, they should be fumigated. Paradichlorobenzine and sulfur are two of the fumigants most commonly used.

Good Queens = A Good Colony

A colony of bees normally consists of one queen (the mother of the hive), thousands of worker bees (sexually undeveloped females) which do all the work in the field and the hive, and some drones (males) whose sole function is to mate with young queens. Since the queen lays all the eggs in the hive, the growth and productivity of the colony is entirely dependent on her. It is therefore highly important that each colony contain a young, prolific queen. The degree of success among large and small beekeepers normally hinges on the amount of attention given to queens. Good queens result in better wintering, faster buildup in population in the spring, less trouble with swarming and a larger crop of honey.

In addition to laying eggs, the queen produces chemical substances which control social order among workers in the hive. Usually the workers respond to the odor of the substances. The presence of such a substance tells the workers that they have a queen; in its absence they begin to produce a new queen. Although workers can lay eggs, they do so only if the queen is absent because she produces a chemical which inhibits their ovary development. A swarm of bees will not leave the hive unless the queen accompanies them. If they are separated from the queen, the bees will return to the place where they were last aware of her presence.

Brood-Rearing and Storage of Honey

Insufficient room for the queen to lay eggs and for the bees to store honey is one of the greatest causes for failure in beekeeping. Swarming is the natural method of reproduction in honeybees. A crowded or congested brood chamber stimulates the colony to swarm. Swarming usually takes place during the period of greatest brood-rearing. This peak is normally in the spring after the dandelions begin to bloom and before the clovers start to yield nectar. During this period it is especially important to provide ample worker comb in which the queen may expand her nest in an upward direction.

The problem of swarm control is greatest between the dandelion and clover honey flows, but it continues later into the season if there is not enough room for the bees to store nectar. Bees short of storage space are forced to hold the nectar in their honey stomachs. When these bees are unable to work in the fields they crowd the brood chamber. This in turn stimulates swarming and results in a reduction in the honey crop. Since bee larvae feeds on honey, any reduction in the honey crop results in a serious loss of life to bee larvae.

Seasonal Management Summary

It is not difficult or time consuming to manage a few colonies for honey production or pollination, but it is imperative to give the colonies adequate attention at definite periods during the active season.

The following summary of seasonal management is designed to help beekeepers manage their colonies efficiently.

The dates given are approximate; in southern counties usually from one week to ten days earlier and in northern counties from one week to ten days later. This plan of operation is satisfactory during an average year, but it should be modified to meet changing honey-flow conditions. In most southern states bees fly the year round.

On the first warm day during late March or early April when the bees are flying:

1. Clean the entrances of the colonies. In the fall, colony entrances must be reduced in size (about 2½ inches long by 3/8 inch high) to help the bees keep warm and to prevent mice from entering. Sometimes dead bees or dislodged insulating materials restrict or clog the entrances and should be moved. As the weather becomes warm in the spring, the entrances should be enlarged.
2. Close the entrances of dead colonies or remove them to a bee-tight building or cellar where bees cannot rob the honey from the combs. This precaution prevents the spread of American foulbrood.
3. Unite queenless colonies or those with failing queens by placing them above strong queen right colonies, separated with a sheet of newspaper. The paper should be pierced in two or three places. If time and temperature permit, it is advisable to remove the failing queens.
4. Strengthen weak colonies containing young prolific queens by placing them above strong colonies with a queen excluder and a piece of newspaper between them. The colonies should be united during the latter part of March or early April and separated toward the end of the dandelion- or fruit-bloom flow. When the colonies have been separated, set the stronger of the two colonies on a new stand. This helps to equalize the field forces of the two colonies because many of the field bees in the stronger colony will return to their old location.
5. Feed the colonies, if necessary, with a mixture of one or two parts of white sugar to one part of water, by weight or measure. Feeding is rarely necessary if ample honey in the comb is provided in the fall.

Late April or early May when the dandelions and fruit start to bloom:

1. Unpack the colonies and check for food, diseases and performance of the queens in egg laying.
2. Provide ample worker comb in which

the queen can lay and freely expand her nest in an upward direction. This is necessary to prevent the bees from preparing to swarm. The brood nest may be expanded in two general ways:

a) Reverse the brood chambers of all strong colonies and add a super of worker comb when the colonies need more room.

b) Add a super of worker comb on top of each colony.

In each method a frame of sealed brood is raised from the lower brood chamber into the super which is added. Both methods provide enough room for the bees to work in an upward direction.

3. Feed if necessary. Starvation of bees between the dandelion and clover flow is one of the principal causes of unprofitable beekeeping. Never let a colony get below the equivalent of three full combs of honey, especially at this critical time of year.

4. Check colonies for queen cells by separating the hive bodies and examining the bottom of the combs of the upper-hive bodies. If queen cells are being built in anticipation of swarming, use one of the following methods to prevent swarming:

a) Remove all queen cells and reverse the brood chamber.

b) Divide the colony into two parts, making certain that the brood, bees and honey are divided about equally. All the queen cells should be removed from the division containing the queen, but one or two of the largest queen cells should be left in the queenless part.

c) Remove all the queen cells and exchange the position of strong colonies in the apiary with weak ones. Enough field bees usually leave the strong colonies, return to their old locations and enter the weak colonies to prevent further trouble from swarming for at least two weeks.

5. Do not clip the wings of the queens if enough room is provided.

6. Fumigate stored combs if even one wax moth larva is found in them.

At the beginning of clover honey flow (usually during the last half of June):

1. Place the queen in the lowest chamber. This is usually accomplished by driving the bees out of the supers into the lowest chamber with smoke, certain acid fumes or shaking them from the combs. Add a queen excluder and a super of drawn combs (super shaken free of bees) and then place the hive body or hive bodies containing brood on top of the super. An additional super of combs should be placed on top of the colony to insure ample space for the storage of honey. Confining the queen to one hive body during the light honey flow is a popular method among beekeepers to prevent unwanted swarming.

2. Allow the queen to lay in two brood chambers throughout the entire season. With this method, swarming is usually prevented by reversing the brood chambers at least once or twice during May and early June. At the beginning of the clover flow, remove five or six frames of brood from each colony and place them in the center of the third super which is added at this time. Shake each frame of brood as it is removed, to dislodge the bees. This permits inspection of the combs for American foulbrood and at the same time prevents the queen from being carried up into the supers. Next, place a queen excluder between the second chamber and the third super which now contains five or six frames of brood. This method provides ample room for the queen to lay in the two lower brood chambers and at the same time stimulates the bees to work in the supers above the excluder.

3. Examine the colonies for room and for queen cells about every two weeks during the clover honey flow.

At the end of the clover honey flow (the latter part of July or early August):

1. Requeen or at least mark the colonies that need new queens so they may be requeened as soon as time permits. Young queens insure good wintering and a maximum honey crop the following year.

2. Examine the colonies for disease before removing the honey crop.

3. Remove and extract the clover honey crop and return the supers to the colonies for buckwheat or other fall honey flows.

At the end of the buckwheat and/or fall honey flow (usually during the latter part of September):

1. Unite all weak and queenless colonies with other strong colonies.

2. Examine the colonies for disease.

3. Remove the supers and queen excluders and reduce each colony to two hive bodies for winter. Make certain that the top hive body or second brood chamber is full of honey.

4. Remove and extract the surplus honey and store the supers for winter.

Bee-Supply Houses

Hives and other beekeeping equipment are manufactured by several companies in various parts of the United States. In the northeastern states the following companies handle a complete line of apiary and honey house equipment including honey containers, package bees and queens. These companies will forward their catalogs on request.

Dadant and Sons, Inc., Hamilton, Illinois

Hubbard Apiaries, Onsted, Michigan

Walter T. Kelley, Clarkson, Kentucky

A. I. Root Co., Medina, Ohio

A. G. Woodman, Grand Rapids, Michigan

Suburban & Farm Catalog
Sears Roebuck and Co.
Dept. 139
4640 Roosevelt Blvd.
Philadelphia, Pennsylvania 19132

Dadant and Sons, Inc., and the A. I. Root Co. have several distributors in New York State as well as in other states. If requested, they will send you the name and address of their nearest distributor or dealer.

10

The Family Forest

INS AND OUTS OF A FAMILY FOREST

One out of every ten families in America owns a "family forest." Since there are more than 20 million farm families in the United States today, millions of acres of prime timberland or/and potentially prime timberland, producing billions of board feet annually, are owned and operated by small family enterprises.

The "small family forest" can be from one to several thousand acres of woodland. Somewhere between those limits is room for the beginning farmer, the inexperienced person.

The family forest has many good uses beyond its value as another source of income for the small farmer. It is a refuge for wildlife, ranging in size from the tiny titmouse to the moose and grizzly bear: When properly managed, it can be your personal contribution to healing the ravages of a technology that is strangling our heritage in a quickening beat of erosion, fire and greedy, heedless exploitation.

It is more. Forests, great and small, are places of refuge for us all; quiet, solitary walks, peace of mind and soul; the forest is a cathedral where we–in our own individual ways–may renew our faith in the quiet strength of nature.

Managing Your Forest

Well-managed forests have certain things in common. The trees are suited to the soil, climate and locality; they produce a salable crop of saw logs, pulpwood and other products. Remove poor or surplus trees to give the good ones room to grow.

Forests can be "even-aged," made up of trees approximately the same age and size, or "uneven-aged," trees of several ages and sizes. The species of trees that make up the stand determine the forest type.

In even-aged stands the crowns of the trees form a ceiling of foliage, or a canopy. The crowns may extend down to one-third or one half the total height of the trees, and the shade retards the establishment and growth of most young trees and other low vegetation. Reproduction may be going on, however, on the forest floor below the main canopy of trees, depending on the forest type and its stage in the cycle of management.

In uneven-aged stands some trees are mature, some just becoming established and some at ages between. The beech-birch-maple combination, a northern hardwood type, is an example of stands usually managed on an uneven-aged basis.

Several of the species common to this forest type are tolerant of shade and become established either under large trees or small openings in the canopy.

The forest floor is covered with needles, leaves, twigs and small branches. Such a covering permits soil to absorb the large amounts of water that trees need, preventing erosion. Usually beneath this litter, a moist, fertile layer of humus covers the soil. On sandy soil the humus may leach away.

The well-managed forest is easily accessible from the outside and contains a system of woods roads within its boundaries for the frequent removal of small quantities of timber products.

Every forest should be operated according to a plan based on (1) the owner's objectives and financial or physical limitations, (2) the size, condition and capabilities of the woodland, and (3) expected markets.

Where the woodland is a part of a farm, management should be integrated with the rest of the farm. Most state forestry departments have "farm" or "service" foresters who work closely with local Soil Conservation Service farm planners, county agricultural agents, and extension foresters in developing the woodland phase of the farm plan. Consult them.

For small forests, especially those under 30 acres, and for woodlands composed entirely of young timber none of which is approaching maturity for the intended products, a simple plan may be all that is needed. The plan would list the action for a period of time, about five years. For example, if reforestation is needed, the plan would indicate the area for planting, number of trees required, species to plant, spacing and approximate planting date. Planting planned for a specific year or for different years would also be indicated. In addition, the plan might show on its map the location of proposed access roads, firebreaks or other improvements. A release cutting (timber stand improvement) may be planned for a certain area, a thinning for another area.

For a larger forest, where some of the timber is mature or approaching maturity and the owner wants a repeated, frequent income, the plan should include a schedule for harvesting the timber crop on a sustained-yield basis. This would require a timber inventory and growth study, based on a program of annual or periodic harvest cutting to assure continual operation without running out of timber. In addition to the development and rehabilitation work, the plan should include a cutting budget showing areas to be cut, approximate amounts to be cut and data on present volume by species and products.

Natural Seeding

The cheapest way to reproduce a stand is through natural seeding. Ease of getting natural reproduction started varies between localities, forest types and species. It depends mostly on the availability of seed, upon available moisture, favorable summer temperature, condition of the

Fire and livestock have been kept out of this thrifty, mixed pine-hardwood forest. The forest floor, rich in humus, will hold maximum rainfall and prevent erosion.

seedbed, number of rodents, insects and diseases, and amount of grazing by domestic and wild animals. In areas where it is difficult to establish natural reproduction, where the summers lack rainfall and the soil is very dry, technical advice should be sought to improve your chances of a satisfactory restocking with desirable species.

Most uneven-aged forests reseed themselves. The kind of trees found in the reproduction following a harvest are often influenced by timing the cut to coincide with a good seed crop of the desirable species and by cutting less desirable species more heavily.

In fully stocked, even-aged stands with a dense canopy, few seedlings will develop into trees because of lack of sunlight. Natural reproduction becomes established when the mineral soil is exposed, sunlight reaches the forest floor and seed is available. If no seed source is present, the area will have to be reforested by other means. This can be done by sowing seeds collected elsewhere.

Planting Trees

The first forest trees planted in this country were wild seedlings dug in the forest and replanted on an unstocked area. Later, seedlings were brought from Europe and planted in parts of New England and the southern Appalachians. Today seedlings are grown by the millions in public and private nurseries. Your local forester can assist you in getting state-grown seedlings without charge.

The time to plant trees varies from place to place and with local climate. In southern areas planting is done throughout the winter when weather permits and the ground is not frozen. In northern areas planting is done mostly in the spring. If the soil is light, or there is no danger of the ground heaving under pressure of winter frost, and there is abundant moisture, fall planting in northern areas after growth has stopped produces good results with some species. If danger of frost heaving exists or if soil moisture is deficient in the fall, planting in early spring is preferable. Planting should always be done before new growth starts in the spring.

Trees for timber production should normally be planted only on land that is better suited for growing trees than other crops.

Many landowners do not wait for natural reproduction after harvesting mature timber, but plant trees on the area immediately following the cut. This is a common practice in certain types of forests and on sites where worthless species of vegetation appear promptly following the harvest. The trees established by prompt planting are better able to compete with the undesirable vegetation than trees resulting from natural seed fall.

It is usually safe to plant the species that are native to the area, but this may not always meet the owner's requirements. Growing conifers where hardwoods are native may sometimes be better. Most state and extension foresters have prepared guides for selecting the species of trees that grow on different sites.

A weeding cut has been made in this young hardwood stand.

Generally conifers will grow well where hardwoods may not: in worn-out fields and pastures; in sandy, burned-over or eroded areas; or in shallow, heavy or cloddy soils. Hardwoods grow well in deep, rather loose, crumbly soils with plenty of room for root development. Most trees, especially the higher-valued hardwoods of the East, need considerable moisture in well-drained soil.

Spacing of seedlings varies with climate, site, exposure, species and market demands. If Christmas trees are to be harvested, spacing as close as 5 by 5 feet can be used. At the other extreme, 10- by 11-foot spacing has been used for fast-growing pines where the market is for lumber only. The usual spacings are 6 by 8 feet or 8 by 8 feet. The general aim is to have six hundred or seven hundred trees per acre when the plantation is established.

Spacing (feet)	Trees per acre (number)
5 x 5	1,742
6 x 6	1,210
6 x 8	907
8 x 8	680
8 x 10	544
10 x 10	435

Planting two or more kinds of trees together in blocks or strips often has its advantages. If one kind is badly damaged by insects or disease, the other kind may grow into good timber.

In block planting, two or more species of trees are each planted in squares of about nine, sixteen or twenty-five trees. Putting in several rows first of one kind of tree and then of another is called strip planting.

Planting stock is designated by two digits separated by a hyphen, the first digit indicating the number of growing seasons the stock was in the seedbed and the second digit indicating the number of seasons the stock was in the transplant bed. Planting stock listed as "2-1" is stock that has been in the seedbed two growing seasons and moved to the transplant bed for one season. In the South, seedlings of certain species grow so rapidly that 1-0 stock is used exclusively. In other sections stock grows so slowly that three or four years are required to produce trees of plantable size. Good planting stock has a good balance between roots and top.

State forest tree nurseries usually produce stock best suited for planting in their respective states. The age of planting stock is not as important as its physical characteristics and condition at time of planting.

Proper care of planting stock from the time it leaves the nursery to the time it is planted cannot be overemphasized. It must always be protected. The roots of many species are extremely delicate and if exposed to sunlight or to hot drying winds for only minutes the trees will not grow when planted. Nor should tree roots be exposed to freezing temperatures.

If the planting site is not too steep and is relatively free from obstacles, a planting machine may be used to advantage. These machines vary in type and capacity, and are designed to meet various conditions. They may be rented from many state forestry organizations, Soil Conservation districts, Chambers of Commerce, and banks.

In most cases the landowner must supply the motive power, usually a farm tractor. He should make sure that he has the right size and design of planter to plant the trees at the required depth. The tractor should be large enough to pull the planter at this depth. The planter should also be equipped with properly designed and functioning scalpers if removal of sod is desirable.

In areas where a mechanical planter cannot operate, a planting bar and planting hoe continue to be the most effective hand tools for tree planting. Shovels, spades and mattocks are also used. The bar is fast and satisfactory for "slit" method planting in light soils with few rocks. It does not work well in heavy soils, or in areas where scalping the sod is desirable before planting a tree.

The basic principles of tree planting apply to all methods of planting—with a mechanical tree planter or planting bar, planting hoe or shovel.

The following rules should be followed to ensure successful tree planting.

1. Where soil moisture is a factor, eliminate competing vegetation (roots and all).
2. Keep roots of trees moist at all times, protected from the hot sun and dry air.
3. Make a hole deep enough for the roots to assume their natural position.
4. Place tree in the hole as near its natural position as possible—roots extending downward and tree at proper depth in the soil.
5. Pack soil firmly around roots so no air spaces adjoin them.

When preparing the planting site, undesirable vegetation (hardwood or brush) should be removed effectively by burning, girdling or bulldozing with special blades, disk plows and harrows, or brush cutters.

Where soil moisture is a critical factor, even sod must be removed before planted trees will live. In sections where summer rainfall is plentiful, the sod may not interfere. Methods of site preparation will vary according to conditions encountered and equipment available. Preparation may consist of removing sod by scalping it from an area 15 by 15 inches with the planting hoe or from a strip on each side of the planting slit with scalpers attached to the planting machine, or it may involve removing brush with a bulldozer. Your local forester can advise you of the most economical and effective method.

For principle 3 a slit or temporary hole can be made with a mechanical planter, planting bar, planting hoe, shovel or any other suitable tool. Holes dug with a mattock or shovel before planting should not be left open because this will permit the soil to dry out. Plant the trees immediately after the holes are prepared. The holes or the slits must be deep enough for the roots of the trees to be placed in their natural position.

Proper placement of the tree in the hole is extremely important. Although the tree may live for several years even if the roots are not properly placed, experience has shown that very often the roots will not grow down but along the surface of the soil.

Helping the Forest to Grow

Many people assume that once a forest is started it will grow by itself. This is true to a certain extent, just as a garden, after being sown, will grow. However, both will grow faster and produce more crops if the operator or gardener protects them from pests, removes the weeds, thins out the plants that are too dense and harvests the crops properly.

Only rarely will the untended garden produce a bumper crop. The same is true of the untended woodland. Most people do not realize the potential of a woodland because in the past so few of them were properly managed, while the results of a well-tended garden are apparent in a few months. It takes years to derive the full benefits of good forest management, especially if the owner must start by planting the trees.

Simply stated, good forest practices consist in keeping all the woodland area producing the maximum high-value products all the time. To do this the stand of trees must be fully stocked but not so dense as to cause overcrowding and not so sparsely stocked that some areas are producing nothing of value.

Most woodlands already contain the potential for future productivity. Young seedlings or saplings of valuable species often are present but suppressed by larger inferior or valueless trees. Seed trees may be present but chance of natural restocking is poor because of a heavy ground cover of weed species. Even pole-sized trees of valuable species often are being suppressed or deformed by overtopping cull trees (poor or diseased trees that have not been cut, "culled," out) left from earlier logging operations.

Timber Stand Improvement or TSI is the term used to identify all cuttings that are not part of the major harvest operation but which are made for the purpose of improving the composition, condition or rate of growth of the stand. There are many different types of operations involved in TSI work, but frequently they are carried out simultaneously as one operation. For ease of explanation, they are described separately here.

Weeding

As the gardener pulls the weeds from among

the vegetables, so the woodland manager, early in the life of the forest, eliminates the weed species of trees. When this operation is done early—while trees are still only saplings, less than 4 inches in diameter at breast height—it is termed a "weeding" or a "cleaning." Rarely is there a market for the weed trees removed, for they are too small. Usually they are cut off at any convenient height and left where they fall. This operation, just as necessary as weeding the garden, is an investment in the forest.

Thinning

Young even-aged stands or groups of trees frequently need thinning when fifteen to thirty years old. Whenever the crowns are crowded or are less than one-third the total height of the tree, the forest should be thinned. Enough trees should be cut so that each crown has room to grow on at least two sides without being crowded for a period of five to ten years.

Thinnings should be made frequently enough to reduce competition for light and moisture, thereby maintaining vigorous growth. The proper stocking for best growth varies greatly in different forest types, sites and geographical areas. Generally about one-quarter to one-third of the wood volume is removed in a thinning. Fast-growing trees can be thinned more than slower growers. Consequently rules of thumb—of which there are many—are poor standards for good forest management. A clump of eight or ten good trees with room on the outside but crowded in the center might be thinned to two or three according to a rule. Wise selection, however, and cutting of perhaps three might give a larger proportion of the group enough room. Likewise, a thrifty young tree should not be cut even if it is growing directly under an older tree that will soon be harvested. In short, each tree should be sized up separately for its chances of growing into a profitable individual. Because of the technical problems involved in a thinning, a forester's advice should be followed.

Liberation Cutting

The removal of trees which overtop seedlings or saplings is often referred to as a "liberation" or "release" cutting. If the overtopping trees are marketable, they can of course be removed by a commercial sale. But very often there is no market for such trees because of the species, poor form or presence of decay. They should then be killed or disposed of in the cheapest way without endangering the desirable trees in the stand. On the basis of species, danger from insects and fire, and other local conditions, the forester can advise the owner which of several methods is best—girdling, felling or poisoning.

Planning a Harvest Cutting

The primary aim of harvest cuttings is to liquidate your investment in mature trees.

Several different methods are applied in making harvest cuttings. The one to use on a specific area depends on the condition of the forest, silvicultural requirements of the tree species, objectives of the owner and volume of timber involved.

Shelterwood Cutting

Shelterwood cutting protects the young trees while they are becoming established. The residual stand is later removed in one or more cutting operations.

In this system, trees to be cut should be marked prior to each operation. The system has these advantages: It can be used with heavy seeded trees such as the oaks, the financial returns are spread over a period of years, and the forest maintains its esthetic values. The system cannot, however, be used with species that demand a great deal of direct sunlight in the seedling stage. Nor should it be used without the advice of your forester.

Protecting the Forest

If the owner of the family forest is to gain the most from his investment he must protect it against damage. Fires, insects, diseases, storms and grazing are some of the enemies of the forest.

The owner, however, may not be able to give his woodland complete protection; some of the destructive agents are beyond his control. Deer and other game cause severe but often unnoticed damage to young forests. Damage caused by wind, sleet storms or wet snow cannot be totally prevented. But the owner can minimize the damage attributed to many causes of loss.

Insects, diseases and fires annually kill some 13 billion board feet of timber, an amount equal to a fourth of the net growth. Many trees not actually killed by these destructive agents may be retarded in growth for several years. Although fire is the most spectacular destructive agent of forests, it ranks behind both insects and diseases in total annual growth loss and mortality. Since 1911 there has been a steady increase in the organized attack aimed at the prevention and the control of forest fires. A similarly effective attack needs to be raised against forest pests.

Fire

Each year uncontrolled fires cost the owners of small forests millions of dollars, yet the cash value of the wood destroyed is only a part of the damage. Burning kills some of the trees and weakens and slows down the growth of others. Bark beetles and diseases enter easily through the burned places. Young trees needed to establish another forest are killed. Fire destroys the fertile, moisture-holding litter on the forest floor, thereby robbing trees of nourishment and exposing the surface to erosion. Fire injuries often lower the sale value of timber products by one half or two-thirds. Woods fires frequently destroy buildings, fences and other improvements; sometimes they take human life.

Cooperation for Protection: In fifty states the State Forester maintains a fire protection system. In some states private forest owners pay part of the costs of organized fire districts. Owners should take the initiative in protecting their woodlands but they will find state and federal organizations eager to help. Forest owners should also inform their farm organizations, local agricultural planning leaders and county agricultural agents of their fire problems.

The local fire warden will be glad to advise the family forest owner on measures he should take to protect his woodland from damage by fire.

Neighbors, of course, can help each other not only in preventing fires but also in detecting and suppressing them. By being alert, especially in dry seasons, they will catch fires when they are small and easy to extinguish.

The woodland owner should know his local state fire warden, have his telephone number posted for emergency calls and promptly notify him when a fire is observed. Anyone who sees a fire should spread the alarm, tell the landowner, neighbors and the warden or other local authorities.

Fire Prevention: Most forest fires are man-made, either intentionally or carelessly; therefore most of them can be prevented.

Ten percent of the farm forest fires are started by owners or their tenants. In some sections of the country the proportion is about one half.

Many owners think that burning the woods will improve grazing, kill destructive insects, drive out snakes and other vermin or improve hunting. Except under special circumstances, the harmful effects of deliberate and repeated burning of the forest usually far outweigh any small benefits. Although carefully controlled burning may be prescribed for the reduction of fire hazard in some areas in the South, for control of brown-spot needle blight and for site preparation for planting and seeding, wholesale burning should not be undertaken except upon the recommendation of a forester.

Cleaning Up Danger Spots: Each forest owner can himself do much to prevent fire. Slash and debris should be pulled away from standing trees. When desirable, as it sometimes is in the West, slash resulting from logging or windthrow should be burned, but only during damp weather and in accordance with state fire laws. The tops and other useless parts of recently killed timber should be safely disposed of and weeds, grass and brush along the edges of fields cut to reduce hazard. Cigarettes tossed by passing motorists or sparks from locomotives may start fires along roads or railway tracks if flammable material is not removed. During long dry spells it may be necessary to prohibit hunting and other forest uses which might add to fire risk.

Firebreaks: Permanent firebreaks are sometimes constructed for the purpose of preventing the spread of fire from one ownership to another or to prevent spread and facilitate control of fires within individual ownerships. The management plan may specify that large areas be divided into 20- or 30-acre blocks separated by firebreaks. These are usually made by plowing or disking to expose mineral soil and eliminate flammable material from a strip 4 to 8 feet wide. Under severe burning conditions, such as caused by high wind, a fire may be expected to jump the firebreak. A break located well ahead of the fire edge provides a safe boundary from which to start a backfire. Firebreaks should periodically be maintained to keep them free of burnable material during critical periods.

(Left) *Snow Creek, Washington*

(Right) *Don't let this happen on your land. Many fires in small woodlands enter from fields or roadsides.*

Putting the Fire Out: Fire-fighting techniques vary greatly, depending upon the severity of burning conditions, topography and other factors. Practices that are suitable under one set of conditions may be entirely unsatisfactory under others. Large fires require specialized equipment and personnel. Recommendations given here concern only small fires.

Generally the small fire should be checked first at its head, the place where it is burning fastest. This can be done by beating it out, making a firebreak in front of it or wetting down the burning material with water. After the head is under control, beat out the fire or make a fireline along the sides and rear. Take advantage of any existing trails, roads or firebreaks that might help stop it. Be sure the fire is out before leaving it.

Tools and Equipment: Anyone who may have to suppress a forest fire should keep adequate tools in a place where they are always immediately available: a hoe or rake to clean firebreaks, an ax for chopping down burning snags, a shovel to throw dirt on the fire and a water bucket for wetting down smouldering embers. In the South many farmers make their own fire swatters for grass fires by fastening a 10- by 18-inch piece of old belting to a hoe handle. Some use a large burlap bag for fighting fires in grass and other light fuels. The state forest fire organizations may be able to place a small supply of tools at a location convenient to several owners. Local fire control organizations in some areas provide fireline plowing equipment for a nominal fee or without cost.

Pests

Insects and diseases affect forest growth in many ways. They kill trees. They weaken tree vitality and retard growth. They deform and stunt trees. They destroy seed and seedlings. The results may be understocking, poor timber quality, encroachment of inferior tree species and even site deterioration.

To repeat, losses from forest pests are today several times greater than those caused by forest fires. Now that the magnitude of the pest problem is being recognized, reduction in losses on a scale similar to those achieved for fire can be expected in the years ahead.

The forestry agencies of most states are organized to work cooperatively with property owners in pest control. The woodland owner should therefore be alert in spotting pest trouble and promptly report to a local official any suspicious conditions observed. If a serious pest outbreak should be developing, the owner may be eligible for state assistance in suppressing it. (Consult your local Conservation Agent.)

Diseases

Diseases that either kill trees, slow their growth or destroy the wood fibers and so make the trees valueless for commercial purposes can be classified into groups based on their habits or their effect on trees.

Heart Rot: Heart rots are the greatest single cause of disease losses in forest stands. These rots are caused by fungi, many of which form conks or fruiting bodies on infected trees.

The growth of heart rot cannot be stopped once it has infected a tree. The only control measure is prevention. The fungus enters a tree at a wound. A top broken off by wind or ice, large limbs broken off, fire scars, logging damage, lightning wounds or any other wound exposing the heartwood of a tree can permit access of the fungus spore. Once established, the fungus causes increasing damage as time progresses.

Cankers: Most cankers of hardwood and softwood trees are caused by fungi. Cankers may resemble mechanical injuries at first, but they remain open and may grow larger while ordinary wounds heal. Hardwood cankers seldom kill the trees. They do deform trees, and the rot that sets in behind them often causes the tree to break at the damaged spot. Severely cankered trees should be removed whenever possible.

Rusts: Most of the important pine cankers are caused by fungi called rusts. These usually produce orange blisters especially noticeable in spring. They often spend part of their lives on one kind of plant, then move to another, a fact that helps in controlling some of the worst of them.

White pine blister rust attacks any of the

five-needled pines and has now spread through much of their range. This rust fungus spends part of its life on pine and part on the leaves of currant or gooseberry bushes. Rust spores formed on these alternate hosts spread to and infect pine in the summer and fall. Blisters develop on the pine in spring, about three years later. Unless infected limbs are removed in time, branch blisters will spread to the trunk and kill the young tree by girdling it. Merchantable trees which become infected can usually be harvested before they die. The usual method of control is to kill or remove the host (currants or gooseberries) from the woodland and from a safety strip around it.

Oaks are not damaged much by rust. On pine the cankers are swollen, spindle-shaped areas on small trunks and branches or large sunken spots on bigger trees. In early spring they form very noticeable orange blisters. Pines one to ten years old are more likely than older ones to be killed by cankers girdling the trunk.

Where infections are serious, landowners should plant or seed the most rust-resistant species. Even in a heavily infected stand many trees can be saved if the cankered branches are pruned before the rust reaches the trunk. The critical time for this is when the trees are three to five years old.

Root Rots: Many root rots cause loss of timber growth, result in windthrow of some trees and kill others outright. Particularly serious is the *Fomes* root rot of conifer plantations in the East. Since the infected part of the tree is not exposed, the disease is usually not discovered until it has been well established and the tree is seriously damaged, dying or blown over.

Dwarf Mistletoe: Dwarf mistletoe is found on nearly all western conifer species. It is a parasite that reproduces by seeds that are too heavy to spread far; the disease is therefore usually transmitted from larger trees to reproduction under or immediately adjacent to the infected tree. Control measures at the time of logging combined with pruning of infected parts of limbs can control the disease in small areas.

Brown Spot: Brown spot of pine needles attacks all southern pines, killing many and preventing others from growing. Small, light gray-green spots first appear on the needles. These change rapidly to brown, encircle the needle and kill the part above. Severe attacks may almost strip the tree each year. The disease is worst in moist seasons and in areas of heavy grass; it is usually associated with longleaf pine types. Since longleaf pine is a fire-resistant species, prescribed burning (fire under complete control during favorable weather conditions and at the proper season) has been found effective. Spot-free seedlings put on new vigorous growth the next season. As soon as the tree reaches a height of 2 to 4 feet the danger from brown spot diminishes.

Insects

Insects attack both healthy and unhealthy trees. Those weakened by fire, grazing, drought, logging damage, overcrowding or disease are particularly vulnerable. The weakened trees serve as breeding grounds for insects which then spread to healthy ones, sometimes killing all trees on vast areas.

The white pine blister rust infecting this tree is easily recognized by the fruiting bodies.

Bark beetles are among the most destructive of the forest insects. Their control is expensive and difficult. Keeping the stand healthy and growing vigorously, preventing accumulations of slash at certain seasons of the year and observing generally good forestry practices will help to prevent bark beetle damage.

Leaf insects rarely kill trees by their first attack. But repeated defoliation may. Even if death does not result, the insects can be a nuisance, especially in woodlands used for recreational purposes. If control is desirable, insecticides will do the job. (Consult your state forester.)

Measuring the Forest and Its Products

Logs intended for lumber or veneer are commonly scaled (measured) in board-foot units. A board foot is equivalent to the volume contained in a board measuring one inch thick and one foot square. The board-foot volume of a log may be obtained by applying any one of several log rules, the choice depending on local customs. Because various log rules have different bases they give different board-foot volumes for logs of a given size.

Measuring Standing Trees

The owner may want to know the number of board feet of lumber that can be sawed from a tree or the volume of pulpwood measured in cords. Tables have been prepared which give approximate volumes in these units, when the diameter at breast height (4½ feet above the ground) and the merchantable height of the tree are known. Because different tree species vary in form and thickness of bark, no general table will serve. The tables give the gross volume. Deductions must be made for defects in the bole of the tree due to rot, cracks, wormholes, shake, and crook or sweep. Your forester will supply tables on request.

Measuring Cordwood

The cord is the cubic unit used to measure various stacked wood products such as fuelwood, pulpwood, excelsior wood and chemical wood. This group of products is commonly referred to as cordwood when so measured. A standard cord is 4 by 4 by 8 feet. Pulpwood is bought in units of various lengths of sticks, 48, 54, 60, and up to 100 inches long. If the wood is excessively crooked or is not compactly piled, deductions are usually made in the gross measurement.

Harvesting the Timber Crop

The woodland owner can sell his standing trees, or "stumpage," or he can harvest them himself and sell the logs and other products.

In the latter case his return will include not only the value of the stumpage but the price of his own labor, equipment and managerial service. He can do the work himself or have it done under contract by an experienced operator.

The owner who does his own harvesting or contracts to have it done can convert his timber to those products that will bring him the most

The wilting leaders on the Eastern white pine are early evidence of attack by white-pine weevil.

profit. Poles or pilings can be separated from saw-log trees, veneer quality logs from saw logs, and pulpwood cut from material not suitable for saw logs.

Owners who plan to sell timber products should locate a market and have a sale agreement before they fell any trees, because cut trees are perishable.

Products of the Forest

Saw Logs: The demand for saw logs is greater than for any other timber product. There are several grades, usually based on the quality of the lumber or timbers they will produce. Structural lumber requires strength; therefore defects affecting strength of the timber are important in determining the grade. Mill lumber is graded on the presence of defects affecting appearance, such as rot, knots, cracks and wormholes. Very low grades of lumber may have local uses or may be sold for firewood.

Usually no saw log under 8 feet in length is accepted. The minimum diameter varies from 8 to 12 inches at the small end of the log, but some mills will accept even smaller logs. A few inches are always included in the log length to allow for trimming the lumber. Be sure to follow local practice for trim allowance or that specified by the purchaser.

Poles and Piling: Poles for electric power transmission lines, telephone communications systems and certain types of buildings are in constant demand, as is piling used to support other structures. A tree of suitable size and form to make one of these products will usually bring more money per unit of board-foot volume than will the same tree cut into saw logs or shorter products. Trees of high quality are required.

Pulpwood: Pulpwood is used for making paper, rayon or other chemical products, pulpboard, wallboards and roofing. Requirements vary widely with the different mills. The forest owner who wishes to cut pulpwood should get specifications from his local forester or from the buyer.

Other Products: Numerous other timber products are important in many localities. Railroad ties are produced from many species of wood, either as center cuts from large logs after the high-grade, side-cut boards have been sawed or as the only product from smaller logs.

Fuelwood and charcoal wood markets exist in certain localities. They utilize low-grade wood of all species, cut to various lengths from one to 4 feet. Fuelwood is often sold by the cord, rick or truckload. A standard cord of dry oak, hickory, beech, maple or longleaf pine provides about as much heat as a ton of anthracite coal or 200 gallons of fuel oil. Dry wood produces much more heat than green wood. By finding a local market for fuelwood or charcoal wood, the owner is often able to dispose of low-quality trees from his forest without out-of-pocket costs (high labor costs eliminate any possibility of profits from low-quality trees).

Other products of the forest include Christmas trees and greens obtained from thinning and pruning young stands, maple sugar and syrup in northern states, ferns from the Northwest, fence posts and rails.

Selling Forest Products

A profitable sale is the payoff in good forest management which determines whether the owner's investments in cash and labor have yielded the highest return practicable.

Too many people sell their trees to the first buyer who offers a lump sum for all their timber as it stands in the woods. Sales of this kind, made without knowledge of the volume, quality and value of the timber, are unwise and often unprofitable. It is of utmost importance that the timber owner know what he has to sell, both in kind and amount. A prime rule of good selling: know how much timber you have for sale.

Finding a Market

Whether the forest owner sells stumpage or converted products, he usually gains by looking over all likely markets before closing a deal. His local forester will know local prices and markets and may be able to recommend firms farther away. It often pays to advertise in the local paper. Of course, the owner will hear about sales his neighbors have made. He can also ask various wood-using concerns for prices.

Timber Sale Contracts

A written contract should be used in the sale of all forest products. It can be prepared by the seller with the help of his attorney, or it may be prepared by the purchaser, who frequently uses a standard form. Timber sale contracts need not be complicated but they should include all items agreed upon.

The timber sale contract should include every item that may be subject to question during the operation. Only a person who is experienced in timber sales can possibly know what these items are. The owner should therefore get the advice of a forester before preparing a stumpage sale contract.

Safety in the Forest

Experience has shown that many accidents occur in the forest—and a large percentage are fatal. The family forest owner, who probably is not accustomed to forest work, is particularly vulnerable, unless he is extremely careful and takes all possible precautions to prevent being injured.

Experienced workers wear hard hats to protect their heads from falling limbs. They wear heavy shoes, trousers without cuffs, and other practical clothing that has proved best for woods work. They know the safe way to carry sharp tools, and use them in a manner that will not result in injury to themselves or co-workers. Experienced woods workers take the necessary precautions to guard against infection by poison oak or poison ivy. They know what to do in case of snake bite. In working with logs, they are always mindful of the possibility of logs rolling or sliding, and they position themselves accordingly.

Some state forestry departments have published safety handbooks; and the U.S. Department of Agriculture, Forest Service, has issued a "Health and Safety Code." When the local forester has reviewed your timber operation with you, he will be able to give valuable recommendations on safety.

SAMPLE TIMBER SALE CONTRACT FORM

The following timber sale contract form is intended only to give prospective sellers an idea of some of the items which should be included. It is not intended for use as a standard contract, for it will not suit the particular set of conditions that prevail in any specific sale area.

______________________________ of _______________________
(I or we) (Name of Purchaser) (Post Office)

_______________, hereinafter called the purchaser, agree to purchase
(State)

from ____________________________ of ____________________
(Seller's name) (Post Office)

_______________, hereinafter called the seller, the designated trees
(State)

from the area described below:

I. Description of Sale Area:

(Describe by legal subdivisions, if surveyed; approximate, if not)

____

____

II. Trees Designated for Cutting:

All ______ trees marked by the seller or his agent with ______ paint spots
(Species) (Color)

below stump height, also dead trees of the same species which are merchantable for __________________ .
(Kind of forest products)

III. Conditions of Sale:

A. The purchaser agrees to the following:

1. To pay the seller the sum of $ _______ for the above-described trees and to make payments in advance of cutting in amounts of at least $ ______________ each.

2. To waive all claim to the above-described trees unless they are cut and removed on or before ________________ (Date), which is the termination date of this contract.

3. To do all in his power to prevent and suppress forest fires on or threatening the sale area.

4. To protect from unnecessary injury young growth and other trees not designated for cutting.

5. To pay the seller for undesignated trees cut, or trees injured through carelessness, at the rate of $ _________ each for trees under 10 inches in diameter inside the bark, measured at stump height, and $ ________ per thousand board feet for trees measuring over 10 inches in diameter inside the bark at stump height.

6. To repair ditches, fences, bridges, roads, trails or other improvements damaged beyond ordinary wear and tear.

II

Raising Livestock on the Farm

For the individual, family or group engaged in diversified or small-scale farming, raising livestock for food and/or added income should be an integral factor of the enterprise.

Before you decide what livestock this will be, get the facts on the initial outlay for the various kinds of animals—how much it will cost to feed and house them, how much time and labor they will require, what equipment and fencing will be needed. Consult your county agricultural agent and competent farmers in the neighborhood, especially if you are inexperienced in handling livestock.

If you want livestock to provide income, you will need to find out what projects have proved profitable in your area and whether there is a market for your products at satisfactory prices.

Look over the buildings, fences and equipment on your place and consider what kind of livestock is best suited to them. If you can use existing facilities, it will save additional investment.

Then decide whether you are going to grow your own feed or buy it. Find out if it can be bought inexpensively from other farmers or if you will have to pay a premium by buying all of it from a feedstore.

Growing your own hay or grain will cut feeding costs considerably, but bear in mind that unless you already own the necessary farm machinery you will have to buy it or hire someone to do the farm work. It is always possible to buy used farm machinery from neighbors or reputable dealers or at one of the many farm auctions held during the slow months in every county of every state and in every province of Canada. Farm auctions are far more than rural "marketplaces" where bargains are commonplace; they are also important social functions and looked forward to eagerly by farm families in both Canada and the United States. Sometimes the cost and use of machinery can be shared with a neighboring farmer.

If family members can do the chores, little or no hired help will be necessary. Remember, though, that livestock must be fed, watered and cared for every day of the year. This restricts family activities to some extent and should be understood at the start.

CHICKENS

Start With a Small Flock

Most small farmers provide facilities for a flock of chickens. The beginning poultryman should start with about one hundred chicks. This number allows for some losses from disease or accident, and will supply enough eggs and meat for a large family.

A still larger flock becomes a source of income if there is a local demand for fresh eggs.

Several breeds of chicken are suitable for small-farm flocks. To begin with, egg production should not be your sole aim. What you will want is a flock to supply both eggs and meat. Rhode Island Reds, New Hampshires and Plymouth Rocks—and crosses of these breeds—are good for meat as well as egg production. Chicks of good quality are usually available from local hatcheries.

One way to start a flock is to buy day-old chicks. Very young chicks require a lot of care, however, and must be kept in a heated brooder house. Or you may prefer to buy older chicks, already started, or pullets almost ready to lay.

Buy only from reputable hatcheries or breeders. Be sure the chickens have been tested for and are free from pullorum and typhoid diseases.

Most commercial hatcheries now have chicks separated by sex, so you can choose pullets for egg production and cockerels for meat.

Feeding Baby Chicks

When you bring newly hatched chicks home, put them in a brooder house. Immediately provide chicks with a starting mash in chick feeders, and plenty of water in drinking fountains. Finely cracked corn can be fed instead of starting mash during the first two days after hatching.

Following this, mash is usually fed as the entire diet until the chicks are four to six weeks old, except for some fine grit that can be mixed with the mash or fed separately in an old tin cake pan set near the edge of the brooder-stove canopy. Allow one inch of space at feeders and one-half inch at drinking fountains for each bird. As the chicks grow, feeding and drinking space must be increased. Keep mash and water available at all times.

Feeding Older Chicks

When chicks are six to eight weeks old, replace starting mash with an all-mash growing diet or a combination of growing mash and grain. Start by adding small amounts of grain and increase gradually until the birds are getting equal parts of mash and grain at about fifteen weeks of age. Grit must be fed when the diet contains whole grain.

Grain is usually cheaper than mash. It has less protein and vitamin value than mash, but this is not a matter of concern—as the birds grow older, they have less need for protein and vitamins.

Grains fed to poultry include corn, wheat, oats and barley. Most poultrymen now use commercially prepared feeds, with quality carefully controlled. Follow manufacturers' directions exactly in feeding these formulas.

If you want to mix your own feed from homegrown grains, get feed formulas and directions for mixing from your county agent or state extension office. These formulas have been tested, so follow directions carefully, especially in adding minute quantities of vitamins and other additives to large batches of feed. Be sure that these additives are stirred into the feed sufficiently to distribute them evenly throughout.

If space is available and the weather favorable, chicks can be put on range (allowed to run free)

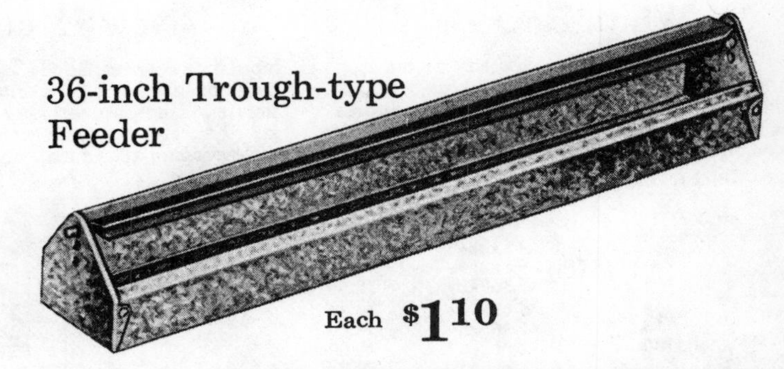

Ample feeding area accommodates 34 fast-growing chicks at one time. Non-tip bottom rests securely on the ground. Metal reel prevents roosting and keeps droppings out of trough. 28-gauge galvanized steel resists rust and fights aging. 3½ inches wide, 1¾ inches deep.
32 KF 89003C—Shipping weight 2 pounds........................Each $1.10

America's most popular Purebred Poultry

Years of research and upgrading by experts bring you high volume, low cost production of superior meat and eggs

White Leghorns

Early to mature . . early to lay . . continue to be good producers over long periods. Small, lightweight breed requires less space, usually requires less feed per dozen eggs yield than larger birds. High quality white eggs bring good prices.

Day-old Chicks—*Postpaid.*

Sex	Box of	Catalog Number	Price per box
As-hatched	25	F32 AF 86251V	$8.50
	50	F32 AF 86252V	15.00
	100	F32 AF 86253V	25.00
95% Pullets	25	F32 AF 86254V	12.95
	50	F32 AF 86255V	24.00
	100	F32 AF 86256V	45.00
*Cockerels	25	F32 AF 86257V	3.50
	50	F32 AF 86258V	5.95
	100	F32 AF 86259V	8.95

Purebred Chicks

Low as 25c each In lots of 100 as-hatched

Superior breeding at moderate costs. Sears purebred chicks inherit excellent egg and meat production qualities from carefully selected parent stock. All breeding flocks have been rigidly culled for egg production characteristics, weight, body type, health, feathering.

All chicks (except Hew Hampshires and Barred Rocks) bred from one of the following superior lines: (1) Wing-banded pedigreed sires from dams with records at rate of 219 or more eggs annually or 180 eggs in 300 days . . or . . (2) sires and dams from a leading breeder and descended from trap-nest pedigree stock with records at rate of 219 or more eggs in 1 year or 180 eggs in 300 days.

New Hampshires and Barred Rocks improved over the years by bloodlines from some of the nation's top breeders. Make your selection from these Sears quality traditional breeds.

Rhode Island Reds

Regular "egg machines," often rival or surpass White Leghorns. No wonder they're among the most popular breeds. Hardy, make an excellent choice in colder climates. Mature early . . grow fast. Fine meat bird; fryers, broilers, capons, roasters.

Day-old Chicks—*Postpaid.*

Sex	Box of	Catalog Number	Price per box
As-hatched	25	F32 AF 86261V	$8.25
	50	F32 AF 86262V	14.50
	100	F32 AF 86263V	25.00
95% Pullets	25	F32 AF 86264V	14.50
	50	F32 AF 86265V	26.00
	100	F32 AF 86266V	48.00
*Cockerels	25	F32 AF 86267V	6.75
	50	F32 AF 86268V	11.50
	100	F32 AF 86269V	19.50

White Rocks

Steady year-round layers of good size brown eggs. Grow quickly into big, deep-breasted birds with outstanding feed economy. Fine meat birds too, dress well with few pinfeathers. Use as fryers, broilers, roasters. Rich yellow skin.

Day-old Chicks—*Postpaid.*

Sex	Box of	Catalog Number	Price per box
As Hatched	25	F32 AF 86271V	$8.25
	50	F32 AF 86272V	14.50
	100	F32 AF 86273V	25.00
95% Pullets	25	F32 AF 86274V	14.50
	50	F32 AF 86275V	26.00
	100	F32 AF 86276V	48.00
*Cockerels	25	F32 AF 86277V	7.25
	50	F32 AF 86278V	12.95
	100	F32 AF 86279V	22.00

New Hampshires

Popular branch of Rhode Island Red family with the same hardiness that makes them good in all climates. High productivity of good size brown eggs. Favorite meat type that's ready for broiler market sooner than many other breeds.

Day-old Chicks—*Postpaid.*

Sex	Box of	Catalog Number	Price per box
As Hatched	25	F32 AF 86281V	$8.25
	50	F32 AF 86282V	14.50
	100	F32 AF 86283V	26.00
95% Pullets	25	F32 AF 86284V	14.50
	50	F32 AF 86285V	26.00
	100	F32 AF 86286V	48.00
*Cockerels	25	F32 AF 86287V	6.75
	50	F32 AF 86288V	11.50
	100	F32 AF 86289V	19.50

Barred Rocks

Develop fast into extra-large, deep-meated birds . . tender flesh, rich yellow skin that dresses out attractively. Good layers too . . start early, maintain high production. One of the oldest and most popular dual-purpose breeds . . keep some for your own use.

Day-old Chicks—*Postpaid.*

Sex	Box of	Catalog Number	Price per box
As Hatched	25	F32 AF 86291V	$8.25
	50	F32 AF 86292V	14.50
	100	F32 AF 86293V	26.00
95% Pullets	25	F32 AF 86294V	14.50
	50	F32 AF 86295V	26.00
	100	F32 AF 86296V	48.00
*Cockerels	25	F32 AF 86297V	7.25
	50	F32 AF 86298V	12.95
	100	F32 AF 86299V	22.00

by the time they are six weeks old. Range gives them exercise and sunshine.

General-purpose birds, such as Rhode Island Reds, start to lay at twenty-two to twenty-six weeks of age. About two weeks before pullets are expected to start laying, gradually replace growing mash with an all-mash laying diet or laying mash with grain.

Feeding Layers

For maximum egg production, laying mash or laying mash and grain should make up the major part of the diet of laying hens. If grain is fed, add grit and oystershell to supply the necessary calcium for normal eggshells. If mash alone is fed, check the manufacturer's directions to see whether limestone or oystershell should be added. Mashes usually contain sufficient calcium.

Feed is the big expense in egg production. Laying hens of the heavier breed all-purpose hens eat 95 to 115 pounds of feed per year.

The average annual egg production of production-bred hens will be two hundred eggs per hen.

Houses and Equipment

Day-old chicks require a well-built, draft-free brooder house that allows at least 1½ square feet of floor space for every two chicks. They will need a brooder stove heated by coal, oil, gas or electricity. Electric brooders are satisfactory and less of a fire risk than other brooders. Some poultrymen use homemade brooders successfully.

The expense and care necessary for day-old chicks can be avoided by purchasing older birds. These may be raised in any building that keeps them dry and protected from the cold, provides ample ventilation in hot weather and permits easy tending of the flock. Such a building can be inexpensive. You may be able to use or remodel an existing farm structure. Allow 3 or 4 square feet of floor space per bird in a barn, unused woodshed or machinery shed, etc.

If roosts are used they should be put at the

Chickenhouse suitable for a small flock. Setting the house on blocks helps prevent dampness and reduces parasites. Other designs of chickenhouses are available from your county agricultural agent.

back of the house, away from drafts. Roosts should be 2 to 3 feet above the floor and 10 to 12 inches apart.

Make a pit under the roosts to catch droppings and to help keep the litter clean. Cover pits with heavy wire netting to keep chickens out of them. Removable droppings boards can also be used. Clean droppings pits and boards often enough to prevent offensive odors.

Cover the floor with 6 to 8 inches of absorbent litter for the chickens to scratch in. Renew litter when it gets damp. Damp litter harbors disease organisms and parasites. Always clean and disinfect quarters and provide fresh litter before housing new birds. The best litter for this purpose is peat moss, but chopped straw and even wood shavings may be used.

Chickens can be confined to houses at all times, or a yard or range may be provided for them. Growing birds, in particular, benefit from exercise, sunshine and fresh air.

Confinement of the flock usually requires less money for land and equipment, and less labor. Losses to predatory birds and animals are also decreased.

Galvanized Steel Nests . . ventilated to help cool eggs, prevent dampne

- Removable bottom trays for easy cleaning
- Sizes for light or heavy breeds

[1] Standard Nest 18^{50} 10-nest size

Roll-out Nest 33^{95} 14-nest size [2]

1 Standard Nests with redwood perches. Embossed ends, partitions allow full ventilation with durable strength. Air circulating through nests helps eliminate dampness. Pinch-grip fasteners cut assembly time in half—squeeze with pliers to join metal parts. Square-hole nests, 28-gauge galvanized steel. Litter guard panels. Bottoms remove for cleaning.

Hinged perches, 1x2 inches wide, swing up to close nests. 60 inches long overall. 10-nest section serves up to 50 heavy-breed hens; 16-nest section serves up to 80 light-breed hens. Your choice of 10-nest or 16-nest section . . order below.

10-nest section. Shipping weight 43 pounds.
32 KF 89035L. $18.50

16-nest section. Shipping weight 50 pounds.
32 KF 89037L. 22.50

2 Front roll-out Nests. Minimize breakage . . eggs roll into individual compartment trays for easy collection . . helps cool eggs and keeps them clean. Trays are made of 1 x ½-inch wire mesh—rust resistant. Designed to prevent sagging, trays are removable for cleaning.

Square-hole nests, built of 28-gauge galvanized steel . . resist rust, fight aging. Durable plastic curtains provide dark interior. 14-nest section handles up to 70 light-breed birds; 10-nest section holds 50 heavy-breed birds. Both 60 inches long. Shipped freight (rail or truck) or express.

10-nest section. Shipping weight 60 pounds.
32 KF 89041N. $31.95

14-nest section. Shipping weight 55 pounds.
32 KF 89042N. $33.95

Artificial Nest Eggs 65c

Enameled finish makes them feel like the real thing. They soothe the hen and often fool her into laying more eggs. Made of quality hardwood. One dozen in carton.
Shipping weight 15 ounces.
32 KF 88171. . . . . Carton 65c

A house for laying hens should provide a nest for every four or five hens in a convenient location. Build sectional nests along the wall in such a way that hens can enter from the rear and eggs can be removed from a door at the front. Shown are some ready-made varieties.

Many poultrymen who keep laying hens confined to the house at all times get excellent production. Artificial lights may be used in the laying house during the fall and winter. This lengthening of the hen's "day" stimulates egg production.

Be careful of crowding which will smother baby chicks. It also reduces the egg production of hens and increases the possibility of disease.

Always keep chicks separated from older birds; this, too, helps protect them from disease.

Watch your flock for signs of disease and act promptly if they occur. Signs may include coughing, sneezing, difficulty in breathing, watery eyes, a sudden drop in feed consumption, droopiness and abnormal droppings.

When disease is suspected, isolate sick birds immediately from the rest of the flock. Get a reliable diagnosis and start treatment at once. Kill very sick birds, and burn the remains or bury them deeply to prevent spreading the disease.

Clean all feeders and waterers regularly. Remove droppings frequently and keep clean litter on the floor. Thoroughly clean and disinfect the entire building at least once a year, and after any sick birds have been held there.

Producing High-Quality Eggs

If you have fed and cared for your laying hens carefully, you should get high-quality eggs for your family or for the market.

Gather eggs from the nests twice daily, and clean and cool them. Eggs should be held at a temperature between 45° and 55° F.

If you are going to sell eggs, do not include undersized or thin-shelled eggs, because size and shell quality affect price.

TURKEYS

Turkeys can be raised satisfactorily on small farms, but they do require special care and equipment. Young turkeys must be kept warm and dry.

Turkeys should not be allowed to run with chickens, and young turkeys should not be kept with older ones. Turkeys should not be put in buildings that have housed chickens within the past three months. Land used for chickens or other turkeys should not be used as range for another flock until at least three years have passed. These precautions are necessary to keep turkeys from contracting blackhead and other serious diseases.

Turkeys are usually marketed as mature roasters or as smaller fryer-roasters, sometimes called broilers. Small-type mature roasting turkeys of both sexes are ready for market at twenty-two to twenty-four weeks of age; large-type at twenty-four to twenty-eight weeks. Large-type hens, however, are often marketed at twenty weeks.

Small-type *white* turkeys make excellent fryer-roasters when marketed at about sixteen weeks. Large-type white females, marketed at thirteen weeks, make satisfactory fryer-roasters.

You should buy about one hundred day-old turkey poults from breeding flocks tested for and free from pullorum-typhoid, typhimurium and sinusitis. Feed and water birds as soon as possible after getting them home.

Feeding Practices

For the first eight weeks poults need a starting mash containing twenty-eight percent protein. After this, they should be fed a growing mash, loose or pelleted, containing twenty to twenty-two percent protein, along with grain—both of your choice.

You may use either commercial feeds or home-mixed feeds based on formulas recommended by state agricultural colleges or the U.S. Department of Agriculture.

Any common grain or combination of grains may be used with the growing mash. If corn is fed, it should be cracked until the birds are sixteen weeks old. If confined turkeys are not fed supplementary green feed, give them a well-balanced growing mash. A less expensive mash

without vitamin supplements may be used when green feed and direct sunshine are freely available.

Houses and Equipment

Poults require a well-built, artificially heated brooder house until they are eight weeks old. Allow one to 1½ square feet of floor space per bird.

Use sand for litter the first two weeks, then add wheat straw or splinter-free shavings. Litter is not needed if the birds are started on a floor with narrow slats three-fourths of an inch apart, or on a floor covered with No. 2½ hardware cloth nailed to removable frames.

Older poults and adult turkeys are best kept in confinement in a well-ventilated building with a dry floor and tight roof. All openings should be screened with heavy wire to exclude small birds and predators. Litter the floor liberally with straw, hay or splinter-free shavings. Add to the litter as required for sanitation. Floor space required by poults up to market age and beyond is about 7 or 8 square feet per bird.

Range rearing is practicable if you have facilities for moving turkeys and equipment to clean ground every two to four weeks during the growing season. If weather is mild, you can start poults on range when they are about eight weeks old. If weather is severe and range shelter is not available, wait until poults are ten to twelve weeks old to put them on range.

The range may be a grass or legume pasture, and should be well-drained and fenced. Roosts and shade should be available. Some kind of portable range shelter on skids generally is needed. Wire walls on the shelter should be strong and close enough to keep out predatory animals—dogs, foxes and skunks. Be careful to latch the shelter door each night after all birds are inside.

DUCKS, GEESE AND SQUABS

One advantage in raising ducks or geese is that they generally require less care and attention than the same number of chickens.

For example, young geese can be put on pasture when they are only a few weeks old. They will need little additional food as long as the grass is green.

Anyone planning to keep waterfowl should obtain additional information on breeds, feeding, management and care from his county agricultural agent or from the U.S. Department of Agriculture.

Squabs are young pigeons twenty-five to thirty days of age. Squabs for the family table or market can often be raised successfully on small farms not suited to chicken raising.

If you plan to sell squabs, first investigate the local market. Squabs usually bring good prices, but the demand is more limited than for chickens and eggs.

Pigeons can be raised in simple, inexpensive houses or in an unused part of a barn or shed. Adult birds feed their young on a substance called pigeon's milk, which is produced in the adult birds' crops.

Each pair of breeders will produce ten to fourteen squabs in a year.

Breeds recommended for producing early-maturing squabs of high-market value are King, Carneau, Mondaine and Giant Homer. Squabs of these breeds should weigh 14 to 24 ounces—a desirable weight and size for an individual serving.

DAIRY COWS

The small-tract farmer often wants to keep a dairy cow to furnish milk, cream and butter for the family.

Owning and keeping a cow on a limited acreage is practicable if ample pasture and hay are available; breeding services are offered in the community; a comfortable, sanitary cow shelter

Young Ayrshire stock in pasture

can be provided; someone in the family has the time and the availability every day to feed, water and milk the cow; and the family can make use of the milk produced.

An average cow, well fed and well cared for, produces enough milk to more than pay for her feed, even if all feed is purchased. She will produce 3,000 to 6,000 quarts of milk per year—more than enough for a family of two adults and three children.

A cow will eat 20 to 25 pounds of hay a day, or three to four tons a year, if no pasture is available. In addition, she will need one to 2 tons of a concentrate grain mix.

Hay costs vary greatly over the country, but usually run from $20 to $60 a ton. The price of concentrates depends on the protein content and may range from $30 to $80 a ton. From 800 to 1,600 pounds of straw are needed for adequate bedding. The average cost of feeding and bedding a cow is $100 to $300 a year.

If part or all of the feed can be grown on the farm, the cost will be reduced proportionately. Generally, 2 acres of good land will provide most of the feed (mainly pasture) for six months of the year and cut feed costs almost in half.

Buying a Cow

Select a cow from one of the five principal dairy breeds—Ayrshire, Brown Swiss, Guernsey, Holstein-Friesian or Jersey.

Jerseys and Guernseys are often used for family cows because they are smaller and do not require as much feed or give as much milk as some of the larger breeds, such as Holstein-Friesian or Brown Swiss. Moreover, the milk of Jerseys and Guernseys is higher in butterfat than that of some of the other breeds.

A cow that is four or five years old and has had her second or third calf is generally a good choice. She will be young enough to have years of production ahead of her, and old enough to have shown her milk-producing ability.

Unless you can use or market a large amount

STROHMEYE
& CARPEN

of milk, there is no reason to pay the high price asked for a heavy milk producer.

The cow you select for family use should be sound and healthy, easy to milk, gentle and free of bad habits.

Examine the cow's udder. It should contain no lumps or hardened tissue, and should have good-sized teats. A large udder does not necessarily mean high milk production; avoid large, meaty udders that do not shrink after milking.

See the cow milked by hand, or better still, milk her yourself a few times. Examine the milk for clots, flakes, strings or blood. To do this, draw several streams of the first milk from each teat on a close-woven black cloth stretched over a tin cup or into a "strip cut" especially designed for examining milk.

Do not buy a cow that kicks, or one that wears a yoke, muzzle or nosepiece. Such devices indicate that the cow has bad habits, such as breaking through fences or self-sucking.

Be sure any cow you buy is free from tuberculosis, brucellosis (Bangs disease or infectious abortion) and leptospirosis. These diseases can be transmitted to man. Make sure the cow has been tested for these diseases by a veterinarian no longer than thirty days before the time you complete the sale.

Summer Feeding

If possible, use approximately 2 acres of the farm for pasture to provide summer grazing for the cow.

Permanent pastures of bluegrass or mixtures of grass slow their growth in the summer and may have to be supplemented to provide a uniform feed supply.

In most of the states across the northern half of the country, alfalfa and ladino clover mixed with grasses produce well during the summer but have to be reseeded every three to five years.

Sudan grass and crosses of sudan grass and sorghum or soybeans make excellent summer pasture in the North. A half acre of this temporary pasture may be planted next to permanent pasture. This provides temporary grazing for the

(Above left) *Brown Swiss*

(Above) *Jersey cattle on white clover and dallis grass*

(Left) *Holstein cow*

cow and the excess may be cut and thrown into the permanent pasture for feed.

Caution: Do not allow cows to eat sudan during its early growth or its regrowth after drought or frost. Sudan grass in these stages may cause prussic acid poisoning.

Do not graze sudan until it is 18 inches high, or cut it for hay until 2 feet high. Tall, yellowish-green sudan is relatively safe, but short dark-green sudan is likely to be dangerous.

In the South, coastal bermuda grass, pearl millet, carpet grass, dallis grass and lespedeza make good summer pasture but do not come on early in the spring. In this region, part of the pasture should be planted to crimson clover or small grains, such as oats, rye, barley or wheat, in the late summer or early fall. This will provide some forage for winter and late spring.

A vegetable garden can furnish a little summer feed. Cows will eat pea vines, sweet cornstalks, cabbage leaves and sweet potato vines.

Winter Feeding

The family cow's winter feed consists of hay and a mixture of concentrates. Alfalfa, soybean, alsike clover or early-cut grass hay are satisfactory. A Jersey or Guernsey cow will need at least 10 pounds of hay a day, and a pound of grain for each 2 to 4 pounds of milk she produces.

A mixture of ground corn and wheat bran is a good concentrate to feed with hay. Some soybean oil meal or linseed oil meal may be added to the diet of hay and grain for extra protein. Or you can buy a reliable ready-mixed feed made for milk cows.

The proportions of hay and concentrate may be adjusted. How this is done depends on the cost of the feeds in your area and how much milk the cow is giving. Sixty-four pounds of concentrate furnish about as much nutritive value as 100 pounds of hay.

Provide a block of mineralized salt in a shel-

tered box for the cow, or add loose salt to her concentrate mix at the rate of one pound to every 100 pounds of feed.

Water the cow at least twice daily in winter and more often in summer.

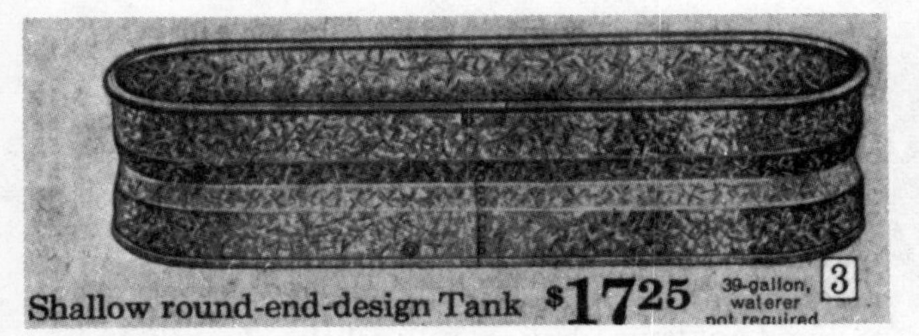

Housing a Cow

The family cow needs a sunny, comfortable shelter or stable. She may be left untied in a box stall 10 feet by 10 feet, or confined to a smaller space and held with a stanchion, chain, rope or strap.

The cow has more freedom but needs about three times as much bedding in a box stall as when she is confined in a smaller space.

If a cow is confined with a stanchion, there should be a manger in front, extending beyond the stall, and a gutter for droppings behind the cow. Allow 4 or 5 feet of space behind the gutter to make it easy for the cow to get into the stall, and to facilitate removal of manure. If possible, have enough space in front of the manger to permit feeding from the front so feed will not have to be carried in from the rear.

If the cow is confined by a stanchion, the sides of the stable should be constructed to prevent drafts in cold weather. This is not so important if she is kept in a box stall. Except in very cold climates, the box stall can be open on the south side if the other three sides are tight. An arrangement that permits the sun to shine into the box stall in winter adds to the cow's comfort. A stall that is entirely enclosed should be ventilated by a tilting window on the side opposite to prevailing winter winds.

Care of Your Cow

Always handle a cow gently and quietly. See that all fences are well constructed so the cow will not develop a habit of breaking through. A fence made of four barbed wires, tightly stretched and fastened to good posts, will keep most cows in a lot.

Brush the cow daily. Do not allow manure to cake on her flanks and thighs. Regular grooming is especially important if she is confined in a stall.

Cows are usually milked twice a day. Before you milk, be sure that the udder and flanks of the cow are free of dirt that might drop into the milk pail. Wash any soiled parts thoroughly. Always wipe the udder and flanks of the cow with a clean, damp cloth before you milk, and be sure your hands are clean and dry. Use a small-top pail and milk with both hands, drawing the milk quickly with as little discomfort to the cow as possible. Keep your fingernails short. Do not insert milk tubes or straws into the teats.

Consult your county agent about breeding services in your area. Artificial insemination will probably be available. Cows are usually bred to calve and freshen (give milk) at about 12-month intervals. It is a good idea for the cow to be dry for a month or six weeks before she calves again. This results in greater milk production. Cows can be made to go dry by reducing their feed and gradually discontinuing milking.

Cows do not usually have much difficulty at calving time, but progress of labor should be checked frequently. If difficult labor is prolonged for several hours, a veterinarian should be called. In cold weather the cow needs a warm stall and plenty of bedding at calving time.

You may want to sell the calf or keep a heifer calf for future milking. Newly born calves of the dairy breeds bring $10 to $30, the amount depending on their size and the current market price. At two or three months, a fattened calf

Young Guernsey grazing

should bring considerably more. If your family does not need all the milk the fresh cow produces, the surplus can be fed to the calf.

Take Care of the Milk

As soon as the milk is drawn, strain it through a clean cloth. Single-use strainer cloths are best. If cloths are to be reused, wash and boil them after each use.

All raw milk should be pasteurized by heating it to 142° F. and holding it at that temperature for 30 minutes, or by heating it to 161° and holding it for 15 seconds. Small electric home pasteurizers cost about $40 at your local feed and grain dealer.

After milk is pasteurized, cool it as rapidly as possible to 50° F. or lower. Keep it in the refrigerator or springhouse until you are ready to use it. Whole milk not needed for immediate use may be held for buttermaking. Keep milk in a deep container until the cream rises to the top. After about 24 hours, skim off the cream for churning. The skim milk may be used on the table, for cooking, or for making cottage cheese.

Rinse all milk utensils in cold water immediately after use. As soon as possible, wash them in hot water containing a dairy washing powder or detergent. Scrub with a brush. Rinse utensils with hot water, then scald them with boiling water.

Keep sanitized utensils uncovered in a clean, airy place. Milk pails and other utensils should be seamless so there will be no crevices in which milk can lodge.

DAIRY GOATS

The small family may find it more convenient and economical to milk one or two goats than to buy milk or keep a dairy cow. Goat's milk can often be tolerated by infants and invalids who are allergic to other milk.

A good dairy goat produces at least 2 quarts of milk daily for eight to ten months of the year and can be fed for about one-sixth the cost of feeding a cow.

A dairy goat costs $35 to $75, depending on her breed and production record. Be sure that any goats you purchase are from a tuberculosis- and brucellosis-free herd.

Feeding

Feed goats producing milk all the clover, alfalfa or mixed hay they will eat, and any available root crops, such as turnips, carrots, beets or parsnips.

If a goat giving milk is not on pasture, a good daily winter ration would be good alfalfa or clover hay, 2 pounds; root crops or silage (chopped green cornstalks), 1½ pounds; concentrates, one or 2 pounds. Concentrate mixtures should consist of oats, bran and linseed oil or other protein supplements.

Goats on pasture need slightly less grain or concentrates.

Pregnant does should be fed all the roughage they will eat in fall or early winter, along with one pound of root crops or silage, and ½ or one pound of the same grain mixture fed to goats in milk.

Any strongly flavored feeds, such as turnips and silage, should be fed after milking so the milk will not be affected by off-flavors.

Keep rock salt before goats at all times and occasionally mix a small quantity of fine salt with the grain mixture. No other minerals are necessary if legume hay (alfalfa, alsike or red clover) is fed. If nonlegume hay such as timothy is used, calcium and phosphorus supplements will be needed. See that goats have access to plenty of fresh water at all times.

Care of Dairy Goats

Goats do not need any special kind of housing, but should be protected from rain, snow and cold. They are natural climbers and, unless tethered, will climb on low buildings and machinery around the farm.

Cleanliness is essential in handling and feeding dairy goats. Does kept in sanitary surroundings do not have objectionable odor. Bucks, the principal offenders, should *not* be kept if breeding service is available in your area.

Milking

A milking stand built with a stanchion at one end and a seat for the milker at one side is a real convenience in milking a goat. Such a stand can be constructed at little cost.

Young does usually object to being milked at first; a stanchion and stand help confine them. A little grain in the box attached to the stanchion helps quiet them. After a young goat is milked a few times in such a stand, she becomes accustomed to milking and will jump on the stand and put her head in the stanchion without assistance.

Heavy-milking does may need to be milked three times a day for a short time after freshening, but twice-a-day milking is usually often enough for grade does.

Breeding and Reproducing

Does come in heat regularly between September and January. After this time, they usually cannot be bred again until late in August. They stay in heat one to two days. The period between heats is generally about twenty-one days. Gestation averages 149 days, or about five months. Does usually give birth to two kids, but occasionally may have three or even four offspring at one kidding.

If your family needs the milk and the farm produces only limited green feed, you will probably want to dispose of the kids. Kids, however, are not hard to raise. They can be fed goat's or cow's milk in a bottle until they learn to drink from a pan, pail or trough.

A good goat will give milk for eight to ten months after freshening. One that gives milk for less than six months should not be kept.

RABBITS

Domestic rabbits are easily raised by small-farm operators; the cash outlay for stock, housing and equipment is modest.

Rabbit meat is all-white, fine grained and high in protein. Only about twenty percent of the dressed carcass is bone, so the meat yield is high compared with that of many meat animals.

In addition to raising rabbits for the home meat supply, you may want to find out whether there is a market in your locality for the meat, or if they are needed as laboratory animals.

All rabbit skins have some market value, especially those from white rabbits because they can be dyed any color.

Selecting Stock

You may buy either young rabbits just weaned or a few animals about ready for breeding. Young rabbits cost less, but the does are not ready to breed until five or six months of age.

Medium and heavy breeds of rabbits are best suited for home and commercial production of meat. Popular breeds include New Zealand, American, Bevern, Champagne d'Argent, Chinchilla and Flemish Giants.

Obtain starting stock from a reliable breeder who will guarantee it as healthy and productive.

The gestation period of rabbits is short—only thirty-one days. A good doe usually raises six to eight young in a litter. She can be bred again when the young rabbits are five or six weeks old.

The most convenient way to milk a dairy goat is from a milking stand as shown here. Note the stanchion and feedbox in front.

This makes it possible for her to produce four or five litters a year.

Young rabbits of the medium breeds are ready to eat or market when they are weaned at two months and weigh about 4 pounds. A good doe can thus be expected to produce more than 100 pounds of marketable rabbits each year.

Feeding Practices

Following is a satisfactory ration for does suckling young:

	Percent of ration
Protein supplement	20.0
Grain	39.5
Roughage	40.0
Salt	0.5

Dry does, bucks and young rabbits do well on the following ration:

	Percent of ration
Protein supplement	8.0
Grain	31.5
Roughage	60.0
Salt	0.5

Dry does, bucks and young rabbits may also be maintained on alfalfa hay alone or hay plus a few ounces of grain daily.

A number of ingredients or combinations of ingredients can be used to meet these nutritive requirements: linseed meal, soybean meal or peanut meal for the protein; corn, oats, barley, wheat or milo for the grain; alfalfa or clover or other good-quality hay for the roughage. Salt can be added to other feeds, or a small piece of compressed salt may be provided for rabbits to lick.

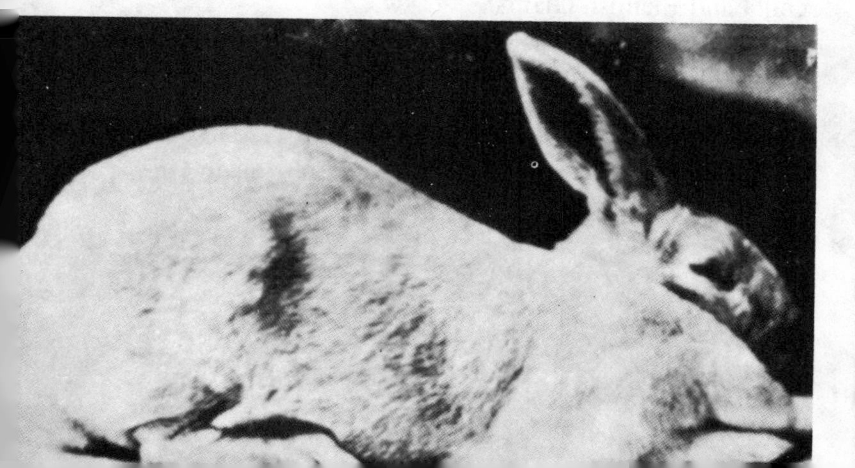

Many rabbit growers prefer to buy a pellet feed—especially prepared for rabbits—that contains all the necessary nutrients and saves the labor of mixing feeds.

But whether the grower uses a pelleted feed or mixes the ration himself, it is important to regulate the feed intake to prevent the rabbits from becoming too fat.

About 400 pounds of grain and other concentrates are required to feed an average-sized doe and her four litters to the age of eight weeks.

Small amounts of green feed—freshly cut grass, clover or garden crops—may be added to the diet.

An inexpensive hutch suitable for small rabbitries. It is light in weight and can be moved from place to place. Hay is kept in the manger between the two compartments.

Housing

Rabbits are usually kept in hutches about 2 feet high, no more than 2½ feet deep and 3 or 4 feet long. These can be made inexpensively at home. Hutch construction varies from all-wire quonset-shaped hutches for use inside buildings to partially enclosed versions for outdoors.

Several types of flooring can be used in building hutches. Wire mesh flooring is used extensively where self-cleaning hutches are desirable. Solid and slat flooring, or a combination of solid flooring at the front and a strip of wire mesh at the back, can also be used.

Flemish giant, a large breed of rabbit. Mature animals weigh 13 to 16 pounds. Varieties differ in color—steel gray, light gray, sandy, blue, white or fawn.

In areas where the climate is mild, hutches can be placed outdoors in the shade of trees or buildings, or under superstructures for protection from sun and rain. Sunlight is not necessary for rabbits. During hot weather, cooling measures must be provided in addition to shade. If buildings are used they should be adequately ventilated.

HOGS

If you raise pigs chiefly on surplus garden produce and table scraps, without having to buy much feed, it will be profitable to raise one or two for the family meat supply.

Buying Hogs

The best time to buy pigs is in the spring when they are being weaned. Be sure that any pig you buy has been raised on clean ground under a strict system of swine sanitation and has been vaccinated for hog cholera.

Choose a female pig, or a male pig that has been castrated (a barrow). A male pig that has not been castrated will produce meat with an undesirable odor and flavor.

Feeding Practices

If hogs have access to good pasture, they thrive on ten to fifteen percent less feed. In the northern half of the United States alfalfa, ladino, red clover, alsike, white clover, bluegrass, burclover, timothy and combinations of these make good pastures for hogs.

In the South, bermuda grass, lespedeza, carpet grass, crabgrass and dallis grass are preferred for hog pasture.

Temporary pasture—rye, oats, wheat, rape, soybeans and cowpeas—can be sown in the hog lot.

A hog that gets no table scraps, garbage or pasture eats about 600 pounds of concentrate ration from weaning age of eight weeks to the time it reaches a weight of 200 pounds. The ration should consist of grains, a protein supplement and a mineral supplement. Corn is the standard grain, but barley, wheat, sorghums or hog millet can also be fed. Yellow corn is more satisfactory than white corn when no pasture crops are fed.

A good mineral mixture consists of equal parts of steamed bonemeal, ground limestone or air-slaked lime, and common salt. This should be kept in a self-feeder where it is available at all times.

Keep an ample supply of fresh drinking water before hogs.

Hogs like root crops—for example, rutabagas and turnips—but these crops are usually less economical to feed than pasture or hay.

A Yorkshire gilt

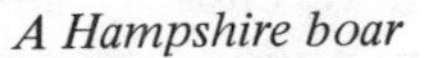

A Hampshire boar

Housing

Any simple shelter that provides protection from drafts, snow and rain, gives shade in hot weather and has a dry floor can be used for hogs. Locate the pen at least 500 feet from any dwelling house to prevent annoyance from odors.

An inexpensive A-Frame hoghouse made with 2-by-4-inch boards and exterior plywood. This portable shelter is big enough for three or four market-sized hogs. A wooden floor can be set in for warmth.

BEEF CATTLE

Raising cattle can put meat on your table or a little money in your pocket more cheaply and easily than any other farming method. You can raise cattle even on a small farm, as long as you have pasture or hayland.

Three of the most popular and least expensive ways of going into the cattle business are these:

1. Feed a few animals for sale or slaughter. You can buy grade steers of a beef breed or dairy steers (castrated bulls) of one of the larger breeds (Holsteins or Brown Swiss).

2. Buy at least three dairy heifers or low-producing milk cows from a dairy herd and use them to raise their own calves and some extras. Breed them to a beef-type bull, such as a registered Angus, either naturally or artificially. Feed bull calves (as steers) for sale or for slaughter to feed your family. Keep heifers for breeding. Buy up to five dairy steer calves for each fresh cow to nurse (fewer for first-calf heifers)—she can feed two sets of two to four calves to weaning during one freshening. Sell steers as grass-fed yearlings. In fifteen years, you can build up a twenty-cow breeding herd while selling yearling steers—thirty from your own cows and three hundred which you buy.

3. Raise a small cow herd of a beef breed and sell weaned calves, yearlings or feeders. If you have use of a neighbor's bull or can use artificial breeding, you can start small. Buy two to five heifers or bred cows. For raising a herd of beef cattle, your farm should be large enough—with enough good pasture and plenty of water—to keep at least twenty-five cows the year round. Unless you have about twenty-five cows, keeping a good bull is usually not profitable. So, if you start with fewer cows, plan to raise the heifers produced in your herd until you get at least twenty-five. You can find out more about such a farm beef herd by asking your county agricultural agent or writing to the U.S. Department of Agriculture, Washington, D.C., 20250. Send your request on a postcard and include your zip code.

What you need

No matter which of the following three ways you choose for your farm, cows are fed and kept much the same way. You will need:

Pasture: Each animal, again, has to have at least 2 acres of good pasture. If your pasture is rocky or covered with brush, 5 or more acres for each animal may be needed.

Hayland: If you plan to keep cattle over the winter, good hay will be needed.

Angus beef cattle

Labor: Raising cattle takes less working time than most other ways of farming. Often, you can do the work yourself after a town job or you can get help from members of your family.

When you have cattle ready to sell, get the buyer to pick them up at your farm or have the cattle hauled to market by a safe, reliable custom trucker.

Buying Your Cows

You can start your breeding herd by buying heifers or cows of beef breeding, dairy heifers, or sound cows culled from a dairy herd because of low production.

If you buy dairy heifers, select Holsteins that weigh 400 to 450 pounds at six months of age. You can raise them on pasture until they are ready for breeding at about eighteen months.

If you buy cows that are culled from a dairy herd, be sure you know why they were culled. You must be very careful in selecting them.

First, you do not want cows that cannot have

Hereford beef cattle

calves. If possible, get cows that are with calf.

Second, you do not want old cows—unless they are cheap. Four or five years is the very oldest you should choose.

Third, you do not want diseased cows. You might catch the disease or the disease might keep the cow from being bred or from having a live calf. Ask for a health certificate signed by a veterinarian that says the cow was tested for tuberculosis, brucellosis and leptospirosis within thirty days of the time you buy.

Either a dairy heifer or a cull cow will cost you $125 to $175.

Feeding

For feeding a 1,000-pound cow and calf to weaning, you can figure on 215 days of pasture and 150 days of winter feed. Winter feed takes 20 pounds of hay and one pound of grain mixture each day.

To fatten a six-month-old steer or heifer from 450 to 1,050 pounds (slaughter weight), you feed for 150 days of winter—8 pounds of hay and 6 pounds of grain each day. Pasture feeding will take care of the next 120 days. You can sell the yearling steer or feed it for 100 days of drylot (off-pasture) finishing—7 pounds of hay and 15 pounds of grain and protein supplement a day.

When you feed grain and supplement, it is easy to mix it yourself. Mix nine parts of barley, corn, oats or other grain with one part of commercial supplement or cottonseed, soybean or linseed meal. A quart of this mix weighs about 1½ pounds.

Put a block of iodized salt in a shady place near water, or add ½ pound of loose salt to each 100 pounds of grain and meal.

If water is not available to your cattle all the time, water them at least twice a day in winter and more often in summer.

Cows need protein and minerals, especially calcium (lime) and phosphorus. Legumes such as alfalfa or lespedeza pasture and hay will give them protein and calcium. Wheat bran, cottonseed meal and soybean meal will give them both phosphorus and protein. Dry cows on good pasture will get everything they need, but you may need to feed fresh cows a mineral supplement.

In the winter, good legume hay and a grain mix that has wheat bran, cottonseed meal or soybean meal will give them the minerals they require.

Winter Feeding

If you wean your calves in the fall, you may be able to feed your herd for several weeks on ungrazed pasture or grain or corn fields after harvest. Because this is not very good feed, you must also give each animal about one pound of high-protein supplement every day.

When the cattle have eaten all the pasture or harvested fields, feed them on hay. As you go into winter, you will need to have one to 2 tons of hay for each cow; cows eat about 2 pounds a day for each 100 pounds they weigh. If your hay is not very good—grass hay that was rained on or straw—add at least one pound of cottonseed meal or other high-protein supplement. You can use 3 pounds of corn silage instead of one pound of hay or one pound of grain mixture for 2 pounds of hay.

Summer Feeding

Pastures are the natural feed for beef cattle. Cattle on good pasture usually will not need anything else to eat.

A pasture of alfalfa or ladino clover mixed with grasses like orchard grass and brome does well during the summer but has to be reseeded every three to five years. Permanent pasture of bluegrass or a mixture of grasses may not be enough during drought.

If pastures are eaten down short or you have a long dry spell, you must feed your cattle some added hay, green corn or other forage.

Feeding grain or protein supplements in May, June or July usually does not pay. If you have it to feed, add grain and protein supplement to pasture feeding in the last 2½ months before slaughter. This way, the meat will be in better condition, will grade higher and will bring a better price.

If steers and heifers are in especially good

shape after wintering, it will be profitable to feed them grain mixture while they are on pasture. Then they can be finished for selling or slaughtering when they are about eighteen months old.

Shelter

Beef cattle do not need much shelter. In almost every U.S. state older animals can find enough shelter in hollows and woods. But during the winter, weaned calves and cows calving do need at least an open shed.

Breeding

Many cows drop their calves in the summer and fall, but most cattlemen prefer to breed their cows for spring calving. Calves dropped in the spring get a better start. The weather is milder and spring calves get more milk than fall calves, provided the herd is on good pasture.

Also, you can sell the spring calves at weaning time and save the expense of wintering them. Or hold them over one winter and sell them the following spring or fall.

Ask your county agricultural agent about how to get your cows settled (impregnated). Artificial breeding costs $5 to $10 for each cow. Tell the technician to use semen from a beef-breed sire.

Cows need a four- to eight-week rest or dry period before they calve. As most cattlemen breed cows in the second to fourth month after calving and it takes a cow nine months to drop a calf, this gives her a calf every eleven to thirteen months.

Most cows that are not with calf come into heat every eighteen to twenty-two days. Unless you have a bull running with your herd, you have to know when each cow is in heat, because this is the time she can be bred. When she is in heat she is jumpy and bothered, bawls a lot and stands to be mounted. Heat lasts 10 to 24 hours. If you can get a bull, use him on the day you see she is in heat. If you use artificial breeding, ask your local technician when to breed her.

Nearly one out of three cows does not settle the first time. If your cow doesn't settle, she will come in heat again in eighteen to twenty-two days and you can breed her again.

Dehorning and Castrating

Dehorn calves. You can do this most easily before the calves are three weeks old. At that age the tender horn "buttons" first appear. Scrape them with a knife and carefully apply the slightly moist tip of a caustic pencil (stick of potassium hydroxide). The caustic causes a scab to form on the irritated area. After a few days the scab shrivels and falls off, leaving a hornless or "polled" head.

Commercial liquid and paste preparations may be easier to use than the caustic sticks. You can also dehorn young calves by applying a heated iron to the base of the horn button.

Male calves must be castrated. Do this at a time of year when flies are absent and before the calves are three to four months old.

You can get directions for dehorning, castrating and marketing cattle from your county agricultural agent.

Weaning Calves

At five to seven months, wean calves that have been running with their dams on pasture. Beginning about a month before weaning, offer the calves around a pound of grain mix each day—to teach them to eat and to lighten the load on the cows. At weaning, take them away from the cows and put them in a pen or barn out of sight of their dams and other cattle and, if possible, out of hearing distance.

Offer the calves some good hay and a small amount of grain at this time. They will eat within a few days. Be sure they have plenty of fresh, clean water in front of them at all times.

12

Raising Feed Crops for Livestock

Left to right: pop corn, sweet corn, flour corn, flint corn, dent corn and pod corn

Like every other chapter in the book, this is written for the small, noncommercial farmer whose goal is to grow enough feed for a few domestic animals on the farm and to leave a late fall surplus in the soil—a surplus large enough to furnish feed and shelter for the wild game that might otherwise starve to death during the long winter months.

The Growing Season

Factors of climate that strongly influence field crop production include the length, temperature and rainfall of the growing season. Temperature is, in turn, affected by elevation and latitude.

Elevation: Field crops are grown from near sea level to approximately 2,300 feet above sea level in hilly areas. Within a region, high elevation generally means a shorter, cooler growing season. A difference of 1,000 feet often makes a difference of two weeks or more in effective growing season for farms that are only a few miles apart.

Summer Temperature: Temperature influences the rate of crop growth. There are important differences among crops. Corn grows best near maximum temperatures in areas where moisture is adequate; moderate temperatures favor oats and grass meadows. The mean July temperature in hilly and mountainous areas in northern zones ranges from a low of 64 degrees in high areas to 74 degrees in the valleys. The maximum temperatures average several degrees higher in valleys and plains than on hilltops 1,000 feet above sea level, where it is impractical to grow most grains. Consult county agent for mean July temperature in your area.

Rainfall: The rainfall during the growing season in farmed areas throughout the nation ranges from far below 15 inches to far above 21 inches. Low summer rainfall increases the likelihood of drought. High summer rainfall favors high yields but often means difficulty in timing harvest operations, especially in the case of hay.

CORN

Seeding

The typical seedbed tor corn is prepared by plowing one to four weeks before planting. Fall plowing is common on heavy clay soils. After plowing, fields are usually harrowed two to four times with a disk or spring tooth harrow.

Planting

Planting begins about May in the northeastern parts of the United States, and early February in the deep South, with the average harvesting season coming five months later. Since 1950 the almost universal use of seed treated with a fungicide has advanced the planting date by a week to ten days. Corn is planted 1½ to 2 inches deep. Kernels are spaced 7 to 9 inches apart in fields which have 36-inch rows in order to provide 16,000 to 18,000 plants assuming seventy to eighty percent of the live seeds become plants. Row spacing varies from 30 to 44 inches and has little effect on yield if the kernel spacing is adjusted to give the desired number of plants.

Fertilizing

Recommended fertilizer at planting time is a 1-1-1 ratio; where extra nitrogen is supplied before plowing or as a side-dressing the ratio is 1-2-2. Manure or a legume sod plowed under can substitute for perhaps one half of the starter fertilizer and all or most of the supplemental nitrogen. A moderate row application is recommended to promote early growth. The ideal placement of fertilizer is slightly to the side and below the seed. Natural manures, compost and the planting of legumes (such as alfalfa or clover) approximately one year before, followed by "plowing under or disking under" just before corn planting time, is recommended for small crop acreages in place of chemical fertilizers. However, the compost pile may be supple-

mented with such recycling materials as leather findings, newsprint, rock dust, bonemeal and literally scores of other materials if natural manure is in short supply.

Controlling Weeds

Weeds rob farmers of many tons of silage and bushels of grain. Early chemical spray plus crop cultivation later make a good team. But chemicals are advisable only where large stands of grains or grasses are planted. Consult your county farm agent.

Again, for the farmer who plants only grains and grasses to feed enough livestock for his own needs, the weed control solution may be "organic" (natural control) rather than "nonorganic" (chemical control). The simplest way to control weeds until the growth of the corn crop has outstripped the need is to lay sheets of moisture-absorbing building paper (or even newspaper) between the rows of corn. The weeds that sprout *with the seed* should be hand-pulled or chopped out with a hand hoe or cultivator. Once the corn has attained a growth of several inches, building paper may be laid down over the rows themselves. Simply cut holes in the paper so that the corn stalks may continue their natural growth freely. Do not remove the building paper after harvesting; on the contrary, plow it under in the spring in order to contribute to the natural composting and enrichment of the soil.

Note: Corn, like cotton, robs the soil of its natural and/or man-made richness. Never follow one corn crop with a second on the same acreage the following season, but alternate it with another grain after the first harvest season.

Storing

Most commercially grown corn for feeding large dairy herds is cut while still green, chopped into silage and stored in upright silos. Ear corn is handpicked from the stalks in the late fall, *husked* and stored in a corn crib. Cribs should be no more than 4½ feet in width. In cribs with these dimensions ear corn will not spoil if cribbed with thirty to thirty-five percent moisture in the grain. The corn is husked and the grain is fed to the livestock over the winter.

Farmers producing 100 bushels or more of shelled corn are no longer as concerned with growing large ears of corn. The development of earlier maturing hybrids and research on plant population and fertility have shown that it is the number of well-matured, uniformly filled ears, averaging ½ pound apiece, rather than the size of each, that produces maximum yields.

The corn plant is a vigorous feeder on plant nutrients and requires abundant moisture to produce maximum yields. Only if soil conditions are ideal and management practices optimum can the corn grower expect maximum yields.

OATS

Planting Oats

The field is commonly plowed in the fall immediately after harvest to facilitate early planting and to create a firm seedbed for a forage crop. Oats should be planted as soon in the spring as a seedbed can be prepared, which may be early April on well-drained soils or late June on poorly drained soils in a wet spring. Oats are drilled (planted) with a seed and fertilizer applicator which can be rented from a neighboring farmer or from your feed and grain dealer, about 1½ inches deep in 7- to 9-inch rows. The planting rate is 1½ to 2 bushels of seed to the acre.

Harvest

In the early years oats were cut by hand with a cradle (a scythe with attached frame to lay the grain evenly as it is cut). Today they are cut with a binder, which cuts, bundles and ties the oats, then shocked or stood upright to dry before being threshed with a mechanical thresher owned and operated by a local contractor who specializes in threshing grains. Only very large farmers can afford to own such expensive equipment.

Storing and Using

Oats for grain are stored on the farm where produced, ground and mixed with other home-

A field of oats

grown or purchased grain and fed to dairy cattle or other livestock. The stalks are used for animal bedding.

SOYBEANS

Soybeans tolerate a wide range of soil moisture. They seldom respond well to direct applications of fertilizer and are very sensitive to fertilizer injury. Plant them about a week after corn at a rate of 3 to 4 pecks in 12 inch rows or 1½ to 2 bushels in rows 7 inches apart.

Soybeans are a grain. They are sometimes mixed with other homegrown grains and ground. When commercially grown they are often processed and the oil extracted for cooking and industrial uses.

HAY (GENERAL)

Hay is the most important crop for any farmer who owns milk-producing animals or raises horses.

Soils

The important forage legumes and grasses have different tolerances and requirements for soil drainage.

Good Hay Soil: Plow layer is dry enough to work most years by the time grasses begin to grow. Subsoil is rarely too wet during the growing season. No mottling. Soils that are "mottled" have yellow, orange and gray or "rusty" streaks, indicating that they are sometimes quite wet during the growing season. Some soils will hardpan (too dense for roots to penetrate) below 24 inches; others have no limit to root penetration.

Average: Spring work is delayed only slightly. About 12 to 20 inches of well-drained soil. Faint mottling below 15 to 20 inches.

Somewhat Poor: Stays wet until late spring.

Strongly mottled below 6 to 12 inches, showing that subsoil is often wet. Sometimes has hardpan at 15 to 18 inches.

Poor: Too wet to work in regular crop rotation. Usually level. Plow layer is dark gray with strongly mottled grayish subsoil.

Legumes, pasture and hay meadow grasses for livestock feed, also differ in their tolerance of soil acidity and response to fertilizer, especially potassium. As a rule these factors should be corrected in order to permit growing the best legume suited to the soil drainage situation.

Climate

Like oats, the hay crops are less limited by elevation and latitude than are wheat and corn. On soils with similar drainage and fertility, the first and second cutting yields of hay are about the same in most areas. In areas with longer growing seasons, legumes like alfalfa may be cut a third time and all meadows grazed later in the fall than in the shorter-season areas. Where winters are very severe, slightly hardier varieties of alfalfa are recommended and more careful management is encouraged.

Planting

At least ninety-five percent of the hay and pasture seedings are made with a companion crop of small grain or, in a few cases, sudan grass. The seedbed must be fine, and for spring seedings the seed tubes of the planter or drill should be set behind the grain tubes. More precise placement can be obtained with band seeding attachments on the drill.

Liming

Acid soil limits the growth of legumes more often than any other factor. Soil pH tests can be made with a simple kit; there is no reason for a farmer to seed without knowing whether the soil is adequately limed. Lime should be considered

an investment and enough applied at one time to neutralize the acidity of the plow layer so that only maintenance applications are needed thereafter.

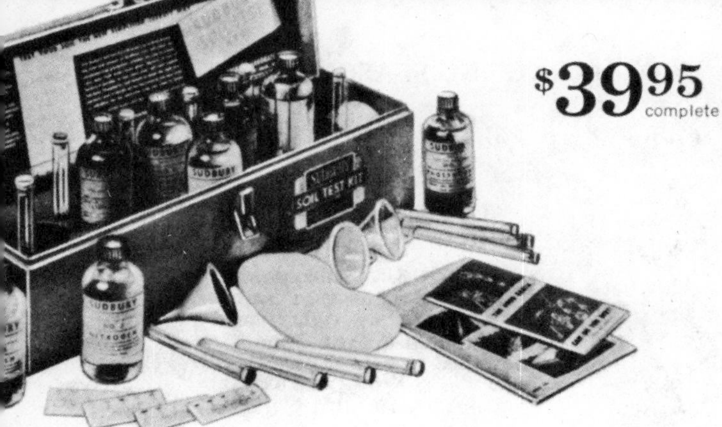

Harvesting

Early cutting is the first essential to high-quality hay. This increases digestibility and palatability, both necessary to obtain a high proportion of feed nutrients from homegrown roughage. Early cutting makes possible aftermath (second growth) and produces it at a time when permanent pastures dry up.

Storing

The hay crop may be stored in the hay mow (storage area under roof of barn) as field-cured hay or in the open, where it is easily accessible for feeding purposes. If stored outside, cover with a large canvas tarpaulin.

RED CLOVER

Before 1955 more red clover seed was used each year than any other legume. Throughout the history of agriculture, it has been the heart of seeding mixtures on most farmland.

Red clover is adapted to a wide range of soil drainage conditions and is easily established, but unfortunately it is essentially a one-year crop. Because of its short life, it does not need to be top-dressed with potassium as in the case of perennial legumes.

There are two types of red clover, medium and mammoth. Medium is preferred as hay because it produces about an equal yield of higher-quality hay. Mammoth is cut only once and is large, coarse and stemmy. The medium type produces a second cutting for hay or pasture. Mammoth is good only as a green-manure soil-improving crop. From bud stage to half-bloom is the best time to cut for hay. Earlier cutting improves quality but some sacrifice in yield will result.

Red clover is the first or second choice as a legume for fields to be cut only one year, then plowed.

ALSIKE

For many years alsike was second to red clover in popularity and is especially suited to poor drainage sites. It is the legume least sensitive to acid soils, but the same liming program is recommended as for red clover. Alsike produces only one cutting in the first year and none the next. It has many hard seeds and reestablishes frequently on wet soils. Stems are weak and it is not drought resistant, but the hay is of high quality. The seeds are very small and are usually planted at 2 pounds per acre in mixture with other legumes.

WILD WHITE CLOVER

This is the smallest of the forage legumes and thus can easily be killed by shading of taller species. Wild white clover is used only in long-term pasture mixtures. It is better suited than other legumes to close, continuous grazing. Sensitive to drought, it makes most of its growth in May and early June except when plenty of moisture is available.

Most native pastures contain enough seeds to

produce a full stand without seeding within two or three years after the pasture is limed and fertilized. Wild white clover has a creeping growth habit. It is seeded at one to 2 pounds per acre in mixtures.

GRASSES

At least one half of the hay and pasture forage eaten by livestock is grass, which is an important part of nearly all meadows after the first harvest year.

Grasses are less sensitive to soil acidity and poor drainage than are legumes. They are also not as sensitive to fall cutting or grazing. The first fertilizer need is usually nitrogen. After the legume has thinned to less than twenty to thirty percent of the mixture, it is usually profitable to start alternate-year treatment with nitrogen one year and complete fertilizer or manure the next.

There are three major grasses—timothy, smooth bromegrass and orchard grass.

Timothy

Timothy tolerates a wide range of soil moisture, is easily established, produces a good yield, and when cut at the proper stage (headed to early bloom) is highly palatable. Its main weaknesses are small amounts of second growth, moderate sensitivity to drought, and heavy leaf loss when cutting is delayed. New late-maturing strains have been developed to retain leaves and color better. Timothy does not start especially early in the spring nor recover rapidly after cutting.

Smooth Bromegrass

Bromegrass begins spring growth about the same time as timothy but retains its leaves and palatability somewhat longer. It is slightly more drought resistant than timothy; new strains have been developed that produce more aftermath. Bromegrass has creeping rootstocks, the stand thereby thickening and filling in open areas. It is better suited for pasture than timothy.

Bromegrass is more difficult to seed than other grasses. The seed is large and chaffy. The most common method for seeding is to mix it with spring grains or early planted wheat and to plant the combination about one inch deep. Planting deeper will spoil the crop. The mixture must be agitated to reduce separation of the grain and lighter-weight grass seed.

Bromegrass is especially sensitive to nitrogen supply. It is best seeded with a perennial legume such as alfalfa and liberally top-dressed with nitrogen fertilizer or manure after the legume thins out.

CROP-MARKETING SEASONS

Corn — *August 1 to July 31* for Louisiana, Oklahoma and Texas; *September 1 to August 31* for Missouri, Kansas, Delaware, Maryland, Virginia, North Carolina, South Carolina, Georgia, Florida, Tennessee, Alabama, Mississippi, Arkansas, New Mexico, Arizona and California; *October 1 to September 30* for all other states.

All Hay — *April 1 to March 31* for Arizona; *May 1 to April 30* for Missouri, Kansas, Virginia, North Carolina, South Carolina, Georgia, Florida, Kentucky, Tennessee, Alabama, Mississippi, Arkansas, Louisiana, Oklahoma, Texas, New Mexico, Utah, Nevada and California; *June 1 to May 31* for all other states.

Oats — *May 1 to April 30* for Georgia, Florida, Louisiana, Oklahoma and Texas; *June 1 to May 31* for Missouri, Kansas, Virginia, North Carolina, South Carolina, Kentucky, Tennessee, Alabama, Mississippi, Arkansas and California; *July 1 to June 30* for all other states.

Red Clover Seed *September 1 to August 31* for all states.

Soybeans *September 1 to August 31* for all states.

Sweet Clover Seed *August 1 to July 31* for all states.

Timothy Seed *August 1 to July 31* for all states.

13

The Kitchen Garden

A kitchen garden, well planned and lovingly cared for, can be a source of tremendous satisfaction for the gardener, beginner or experienced, as well as a continuous source of fresh, nutritious vegetables the year round. Perhaps most important, a well-planned vegetable garden will save the family a great deal of money on the food budget, and provide vegetables with a flavor never found in the bins of the local supermarket. If you wish, the surplus can be sold at your own roadside stand, such as the typical one pictured on the next page.

These benefits may be obtained only by the gardener who is willing to work. In the spring he must work the ground and plant the crops. As the crop matures during the late spring and summer, there is weeding, thinning, replanting and harvesting.

Where to Locate Your Garden

Since vegetables thrive best in full sunlight and need anywhere from 4 to 6 hours of exposure to the sun during the middle of the day, an unshaded slope with a southern exposure is the ideal site for your kitchen garden. Excessive shading can only result in rank, spindly plants and minimum yields.

If you can't avoid shade altogether, remember that beans, cabbage, broccoli and the leafy vegetables can tolerate shade better than cucumbers, potatoes, corn, tomatoes, root crops and melons. Trees growing too close to the garden plot, aside from cutting off sunlight, tend to draw off water and nutrients needed by the vegetable crop.

Whenever circumstances allow, locate the garden within a reasonable distance of the house and as close to a dependable supply of water as possible. The area should be adequately fenced in to protect your crop from livestock, deer, woodchucks and rabbits.

How to Plan Your Garden

Whether you are a rank beginner or an old-time garden buff, a garden plan drawn to scale on paper is a valuable aid in getting the most out of your garden plot. The planning should begin in February or early March with the arrival of the new seed catalogs.

If you intend to plant perennial crops such as strawberries, rhubarb and asparagus, plant them at one end of the plot so that your frequent work on the rest of the garden will not interfere with these plants. Although perennial plants seldom produce a worthwhile yield the first year of planting, they are a definite asset to any home garden because, once established, they can be harvested in the early spring when other vegetables are in short supply. Planting perennials is one sure way to extend the harvest period of your garden.

Your garden should be so arranged that tall-growing plants, such as corn and staked tomatoes, are grown together on one side of the garden plot. This will help prevent them from casting too much shade on your lower-growing crops. Most experienced gardeners choose the north side of the garden plot for planting their taller crops.

Plants maturing early in the season, such as lettuce, spinach and radishes, should be grouped together so that after they are harvested, the ground they have been planted in can be reworked, fertilized and replanted to crops that will mature later on. This is another way to extend the harvest period of your garden.

A third way is to plan several plantings of vegetables such as lettuce, radishes and beans. Space the plantings two to three weeks apart in

VENT OPEN

Strawberry, tomato plants, etc., can be raised from seeds in this coldframe, then transplanted to kitchen garden. Consult your county farm agent for types of plants to raise and for transplanting dates in your particular locality.

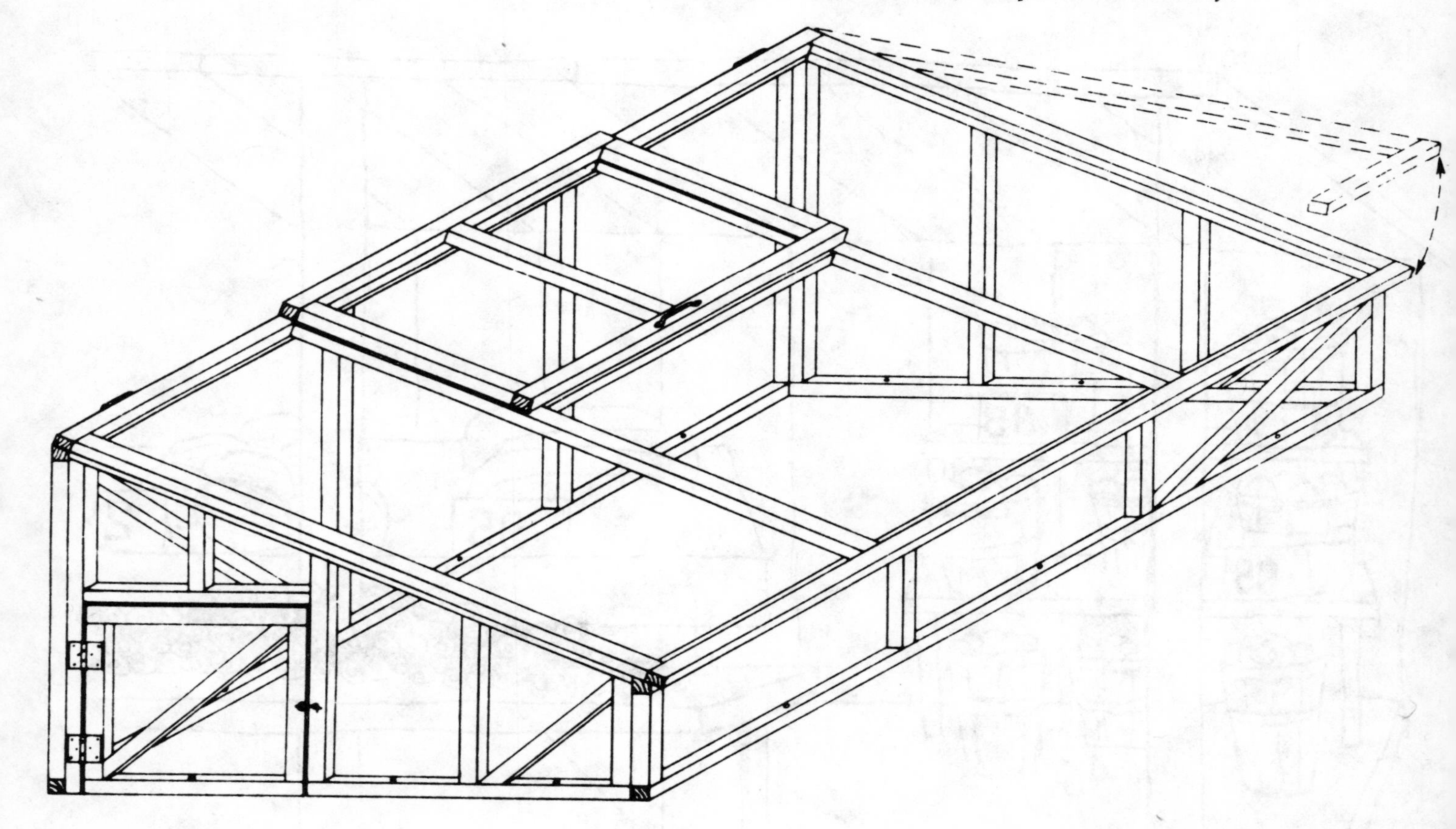

COLDFRAME UNIT

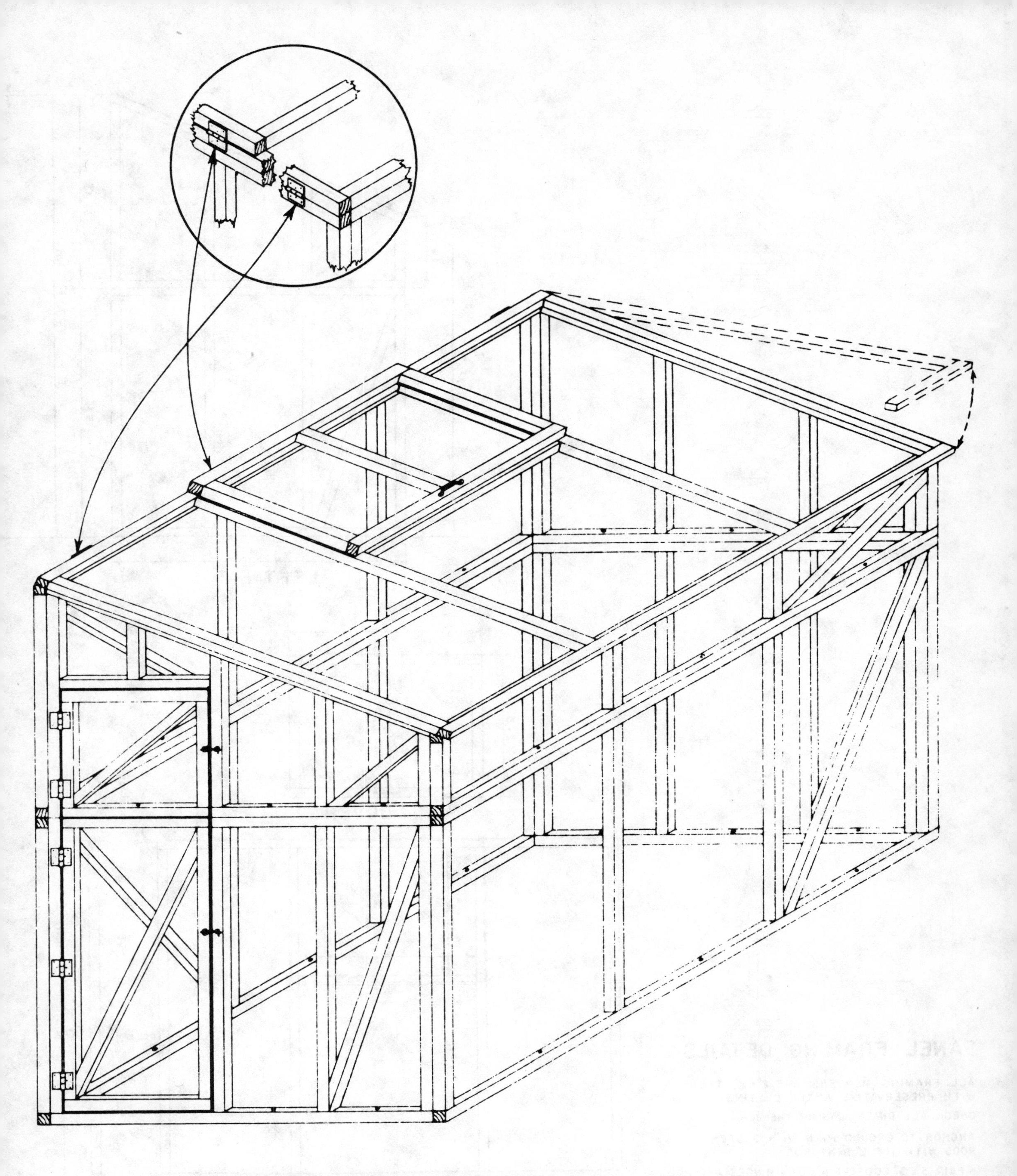

GREENHOUSE ASSEMBLY

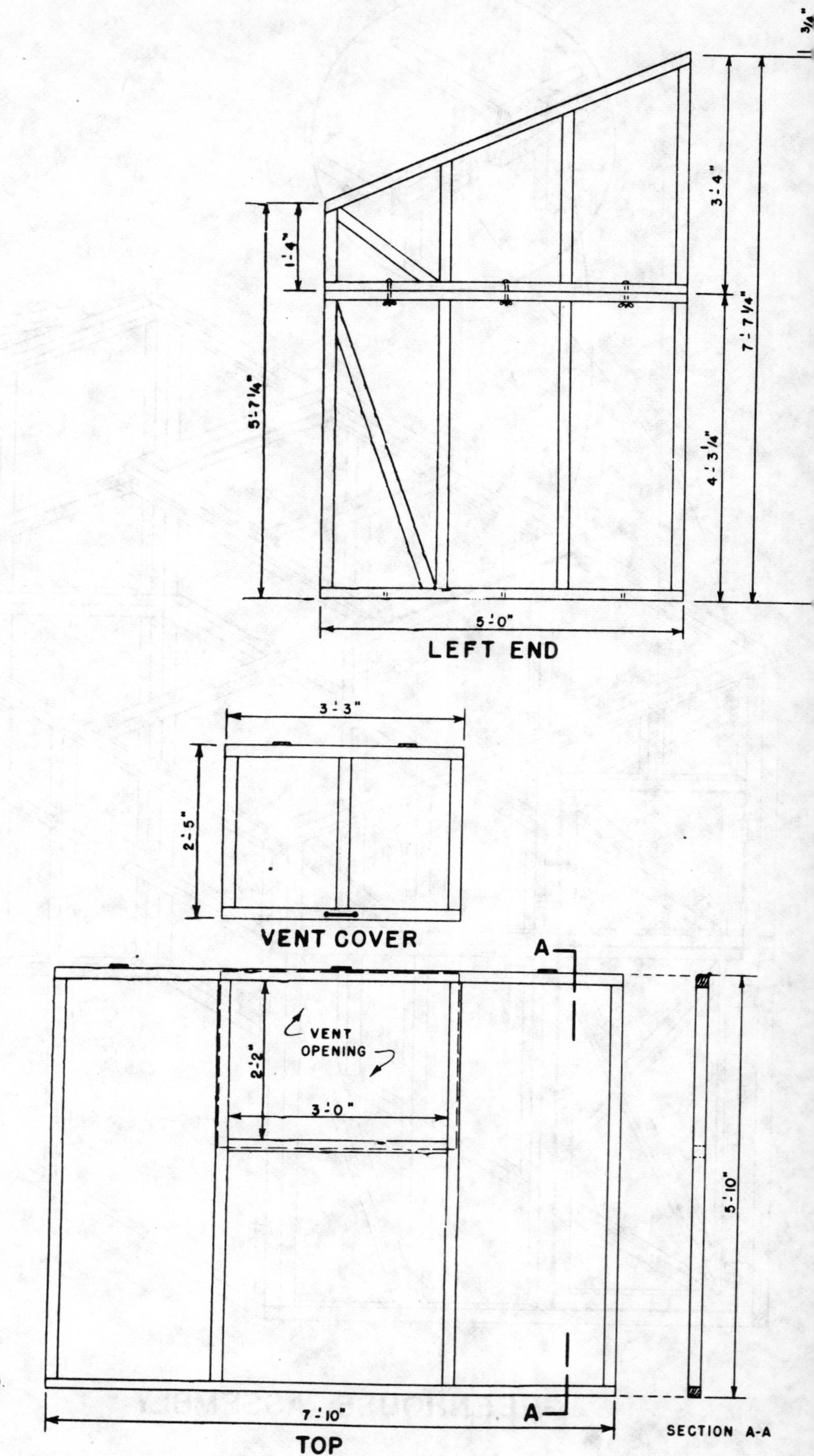

PANEL FRAMING DETAILS

ALL FRAMING MEMBERS ARE 2" x 2", TREATED WITH PRESERVATIVE AFTER CUTTING.

CHECK ALL DIMENSIONS ON THE JOB.

ANCHOR TO GROUND WITH 3/8" x 15" STEEL RODS WITH TOP 2" BENT 90°.

5 PAIR 3" x 3" LOOSE-PIN BUTT HINGES ARE REQ'D.

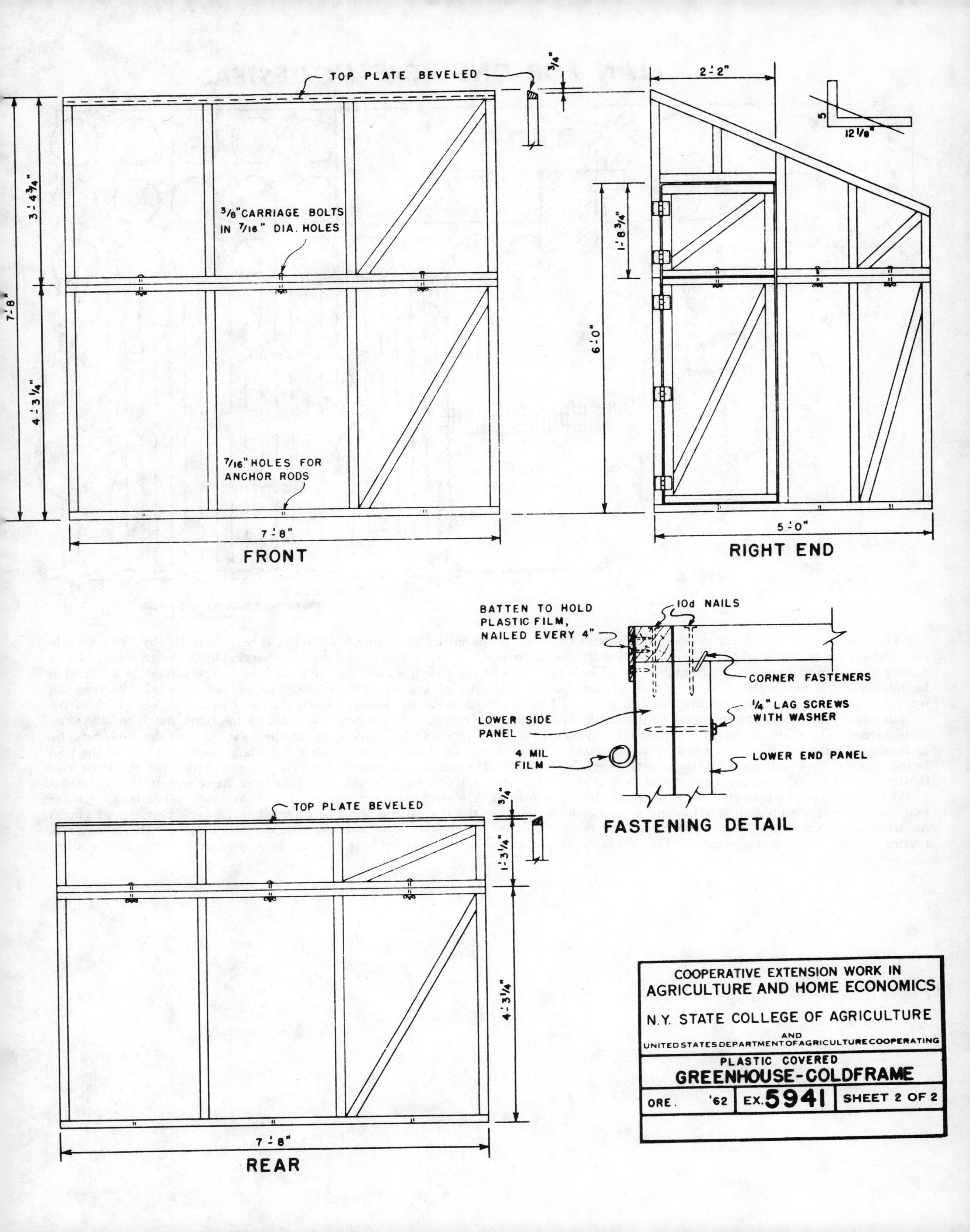
TOP PLATE BEVELED
3/4"
3/8" CARRIAGE BOLTS IN 7/16" DIA. HOLES
7/16" HOLES FOR ANCHOR RODS
3'-4 3/4"
4'-3 1/4"
7'-8"
7'-8"
FRONT
2'-2"
5
12 1/8"
1'-8 3/4"
6'-0"
5'-0"
RIGHT END
BATTEN TO HOLD PLASTIC FILM, NAILED EVERY 4"
10d NAILS
CORNER FASTENERS
1/4" LAG SCREWS WITH WASHER
LOWER SIDE PANEL
4 MIL FILM
LOWER END PANEL
FASTENING DETAIL
TOP PLATE BEVELED
3/4"
1'-3 1/4"
4'-3 1/4"
7'-8"
REAR
COOPERATIVE EXTENSION WORK IN
AGRICULTURE AND HOME ECONOMICS
N.Y. STATE COLLEGE OF AGRICULTURE
AND
UNITED STATES DEPARTMENT OF AGRICULTURE COOPERATING
PLASTIC COVERED
GREENHOUSE-COLDFRAME
ORE. '62
EX. 5941
SHEET 2 OF 2

PLAN FOR ONE ACRE HOMESTEAD

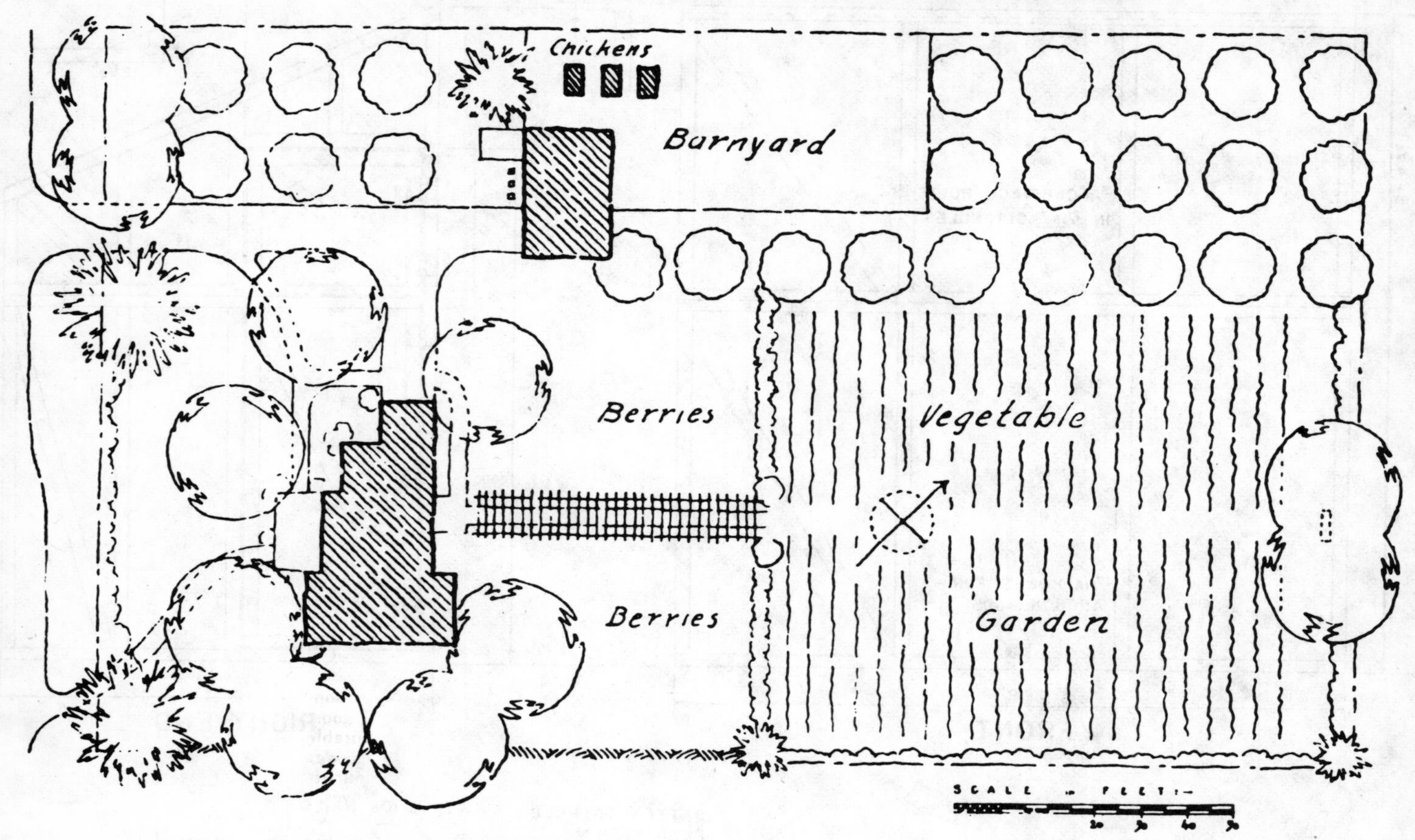

The above suggestion for a one-acre homestead has been prepared by the School of Living. The plot is 160 by 270 feet. The short dimension on the road has been selected with a view to keeping down the road frontage chargeable to the plot and to producing a good ratio between width and length. The longer axis of the plot lies northeast and southwest. The ground is assumed to be relatively flat, an assumption which eliminates special problems of grades or drainage. The prevailing wind directions are from the northwest in winter and southwest in summer. Climatic conditions are assumed to be the same as in the New York metropolitan region. The major ground areas consist of the lawn about the house, the barnyard, the orchard, the berry patch and the vegetable garden. These are obviously interrelated and located so that one will be convenient to the other and to the buildings; the orchard and berry patch are discussed below. ORCHARD — Fruit trees are shown in rows along the whole northwest side of the plot. In this position they will not shade the vegetable garden and do offer some protection to it. They partly screen out the barn from the road and from the house. They are shown as dwarf varieties for obvious reasons. The two large trees next to the main road might be nut trees. Twenty-five fruit trees are shown, affording a wide variety of apples as well as many other fruits. BERRY PATCH— The space immediately beyond a narrow lawn space at the rear of the house is devoted to flowers, berries and grapes. An arbor is shown for the grapes on the axis of the back door. Flowers occupy both sides of the arbor toward the house, with berries beyond. There is ample area in this section to put grapes on wire supported by posts and plant ornamental vines on the arbor.

VEGETABLE PRODUCTION

The following tables contain detailed information about the model garden recommended by the School of Living for the typical family of 4.5 persons. This garden would provide a liberal and a well balanced diet for the average family throughout the entire year.

Vegetables[1]	Number of Feet of Row[2]	Distance Between Plants[3] Inches	Quantity of Seed[3,4]	Approx. Cost of Seed[4]	Approximate Yield Per Foot of Row[2]	Total Normal Yield Pounds	Yield Adjusted for Crop Failure Pounds[5]
Potatoes, White	500	12	50 lbs.	3.00	1 lb.	537	500
Tomatoes	200	24	1 pkt.	.30	2-2½ lbs.	430	400
Corn	110	12	⅓ lb.	.75	3 ears	107	100
String Beans	270	3	1¾ lb.	1.50	½ lb.	135	125
Lima Beans	54	36	1 pkt.	.35	½ lb.	27	25
Peas	120	2	1 1/5 lb.	1.25	½ lb.	64	60
Carrots	100	2	1/3 oz.	.50	½ lb.	107	100
Onions	75	2	1½ oz.	1.50	1-1½ lbs.	86	80
Beets	55	2	1 oz.	1.25	½-1 lb.	43	40
Turnips	55	8	1 pkt.	.30	1 lb.	54	50
Radishes	60	1	4/5 oz.	.50	⅛-¼ lb.	11	10
Parsnips	10	3	½ pkt.	.25	1 lb.	11	10
Cabbage	85	18	¼ pkt.	.35	1½-2 lbs.	135	125
Lettuce	54	6	2 pkts.	.60	½ lb.	27	25
Spinach and Greens	160	6	3½ oz.	1.50	½ lb.	80	75
Brussels Sprouts	44	18	1/10 pkt.	.75	½ lb.	22	20
Broccoli	44	18	1/7 pkt.	.65	½ lb.	22	20
Kohlrabi	20	6	½ pkt.	.25	½ lb.	11	10
Eggplant	10	24	1/10 pkt.	.50	1 lb.	11	10
Okra	6	15	1/3 pkt.	.25	1 lb.	6	5
Peppers	6	18	1/16 pkt.	.25	1 lb.	6	5
Cucumbers	15	36	¼ pkt.	.30	1/5 bu. per plant	43	40
Squash and Pumpkins	120	72	1 oz.	.55	4 per plant	80	75
Cantaloupes & Muskmelons	30	36	½ pkt.	.30	4 per plant	43	40
Watermelon	15	36	2/3 oz.	.70	4 per plant	64	60
Asparagus	100	18	75 roots	6.00	1 lb. per root	107	100
Rhubarb	50	24	25 roots	10.00	2 lbs. per plant	54	50
Totals				34.40		2323	2160

1 Celery and cauliflower are difficult to grow and are not recommended except to very skillful home gardeners (Cornell Extension Bulletin 344).

2 Based on studies published by Utah State Agricultural College cooperating with the U. S. Department of Agriculture in Circular No. 54, Cornell Extension Bulletin No. 344, and the Ball Blue Book published by Educational Department of Ball Brothers Co., Muncie, Ind.

3 Based on Cornell Extension Bulletin No. 344 and Utah State Agricultural College Circulars No. 54 and 93.

4 Based on prices quoted in a reliable seed catalog. The School of Living recommends raising all plants from seed except the perennials. 1969 prices per packet when stated ½ pkt, etc.

5 Adjusted yield is 7 percent less than normal to allow for crop failures.

FRUIT PRODUCTION

	Number of Trees or Plants[2]	Age at Planting	Size When Planted	Cost Per Tree[3]	Total Cost of Trees	Age When Bearing Begins[4]	Duration of Yield[4]	Yield Per Tree[4]	Conversion into Pounds	Gross Yield in Pounds	Yield Adjusted for Crop Failures[5]
Apples[6]	4	2 yrs.	3 to 4 ft.	3.50	14.00	6 to 8 yrs.	20 to 50 yrs.	6 to 10 bu.	50	1200	960
Peaches[7]	3	1 yr.	2½ to 3½ ft.	4.30	12.90	4 to 5 yrs.	10 to 15 yrs.	3 bu.	60	540	432
Pears[8]	2	2 yrs.	2½ to 3½ ft.	4.00	12.00	6 yrs.	15 to 25 yrs.	1½ bu.	65	192	156
Cherries, sweet[9]	2	2 yrs.	3½ to 4½ ft.	5.85	11.70	5 yrs.	20 to 25 yrs.	1½ bu.	41	124	100
Cherries, sour	1	2 yrs.	3 to 4 ft.	4.55	4.55	5 yrs.	20 to 25 yrs.	1½ bu.	40	60	48
Plums[10]	1	2 yrs.	3 to 4 ft.	5.00	5.00	5 to 6 yrs.	20 to 30 yrs.	2 bu.	56	112	90
Grapes[11]	15	1 yr.	12 in.	.90	13.50	3 yrs.	60 yrs.	15 lbs.	15	225	180
Strawberries[12]	25			.20	5.00	1 yr.	5 yrs.	¾ qt.	1¼	32	26
Blackberries[13]	6		8 to 12 in.	.50	3.00	1 yr.	10 to 12 yrs.	1¼ qts.	2	15	12
Raspberries[14]	10		8 to 12 in.	.60	6.00	1 yr.	10 to 12 yrs.	1 qt.	2	20	16
Gooseberries[15]	3			.55	1.65	1 yr.	10 to 12 yrs.	2 qts.	4	12	10
Dewberries[16]	5			1.55	2.75	1 yr.	10 to 12 yrs.	2 qts.	4	20	16
					92.05						2046

1 The kind of fruit grown will vary somewhat according to the locality of the homestead. Currants and gooseberries are recommended by the School of Living in sections suitable for their growth. The U. S. Department of Agriculture advises against growing these fruits in localities where the white pine flourishes because a very destructive disease of this pine lives and thrives on the bushes. Rhubarb and melons are included in the Bulletin on Vegetable Gardening as the operations involved in raising these fruits are more similar to those used in the care of vegetables.

2 Extensive studies by the School indicate that the number of trees and plants recommended in this table will yield a really healthy supply of fruit for the average family of 4.5 persons.

3 Based on nursery prices. Homemade transplants would be less expensive.

4 Estimates based on adjusted data from U. S. Dept. of Agriculture.

5 A deduction of 20% has been made from the gross yield to adjust for crop failures once in every five years.

6 Recommended varieties are Rhode Island Greenings, Red Astrachans, McIntosh, Delicious, and Stayman Winesap.

7 Recommended varieties are Elberta, J. H. Hale, and Mayflower.

8 Recommended varieties are Bartlett, Seckel, Kieffer, and Bosc.

9 Recommended sweet varieties are Governor Wood, Black Tartarian, and Schmidt. Recommended sour varieties are Montmorency, and Early Richmond.

10 Recommended varieties are Lombard, Albion, Imperial, Epineuse, and Stanley.

11 Recommended varieties are Portland, Fredonia, and Concord.

12 Recommended varieties are Howard 17 (Premier), Catskill, William Belt, and Everbearing (Rockhill and Gem).

13 Recommended varieties are Eldorado, Snyder, and Early Harvest.

14 Recommended varieties are Cuthbert, Newburgh, and Cumberland.

15 Recommended variety is Poorman.

16 Recommended varieties are Lucretia, and Mayes.

order to provide yourself with a continuous supply of these fresh vegetables throughout the summer season.

In February or early March when your thoughts turn to spring and the latest seed catalogs, be sure that the catalogs you send for are from reputable seed dealers. If you've never dealt with seed dealers in the past, ask your nearest neighboring gardeners for their lists. They will be only too happy to comply, as well as help you in any other way they can.

Unless you are an old hand at gardening, begin with the time-tested varieties of vegetables that have proven themselves in your particular area. Next year, with a whole season of experience under your belt, experiment with small plantings of two or three new varieties each year. If they do well, you may then adopt them as standard plantings for the future.

As you leaf through the seed catalogs, you may be confused by the frequently great differences noted in the number of days it takes for different varieties of the *same* vegetable to mature. The explanation is simple enough. Some of the plants have been developed to mature early in the season; others, of the same variety, to mature later on. Again, the end result is to extend your harvest season.

Plant corn in *blocks* at least three to six rows wide, not in single long rows. Thus, if you have decided to plant 80 feet of corn, you should plant eight rows 10 feet long rather than one row 80 feet long or two rows 40 feet long. Block planting facilitates pollination and will give you corn of much better quality.

If your garden is to be planted on sloping ground, plant your crops *across* the slope, not up and down. Otherwise, excess water runoff during rainy periods will cause erosion of the land.

Another thing to remember in planting a vegetable garden is crop rotation. Never plant the same crops in the same place year after year. Crop rotation is essential in avoiding plant diseases. These will eventually build up in the soil with continued planting of the same crop.

Equipment

The gardener's best friend is a roto-spader, as shown, or small garden tractor to eliminate the backbreaking labor of preparing the soil and cultivating the garden by hand. It is cheap to purchase, economical to use, and will do a fine job even on very stony soils.

Other tools you will need include an iron rake for smoothing the soil, a square-ended and pointed hoe for planting and hand cultivating, and four pointed wooden stakes with some heavy cord. The cord should be twice as long as the garden. These last items will help you mark straight rows for planting. A wheel hoe or hand cultivator, a trowel for transplanting and a wheelbarrow for all-around use pretty much completes the list, although you will find yourself adding odd bits of smaller tools as you get into the swing of things.

It stands to reason that in order to get the most out of your tools they should be kept clean from encrusted soil, well sharpened and rust free. Wipe them with a well-oiled rag and store them under cover when they are not in use.

4-horsepower. Has single speed, meets same tilling needs as model at right but has sixteen big 12-in. diameter chisel tines for deeper, smoother digging. Cuts a swath up to 26 in. wide, and 9½ in. deep. Welded steel chassis and cast-iron transmission. 4-cycle CRAFTSMAN® engine with compression release, recoil starter. Durable worm-gear drive sealed in oil for more dependable operation with less maintenance. Belt drive with idler.
Fixed drag stake adjusts in height. Adjustable handle raises or lowers for individual comfort. 10-inch plastic wheels adjust in height. Dust shield protects engine. Shipped partly assembled. Red and white.
32KF29003N-Shipping weight 122 lbs. **$134.88**

3½ horsepower. Designed for average-siz lots and gardens. Twelve 11-inch chisel tine cut a swath 22 inches wide and up to 9 inche deep. Swath adjusts to 11 in. wide by sim ply pulling spring clip pins . . no tools neede
CRAFTSMAN® 4-cycle engine with com pression release for 50% easier starts. Reco starter. Rugged cast-iron transmission. Welc ed steel chassis. New improved drag stake fc better depth and tilling control. Plastic 7¼ inch diameter wheels adjust in height fc depth control. All controls convenientl located on central panel. Adjustable handl Deck-type shield helps protect engine fro dust, dirt. Shpd. partly assembled. Red, whit
32KF29102N-Shipping weight 114 lbs. **$119.8**

What Size Garden Should You Plant?

A garden 50 feet wide and 100 feet in length will produce a bountiful harvest of such important perennials as asparagus, rhubarb and strawberries. It will also supply the family with plenty of fresh vegetables during the growing season and produce a surplus of crops large enough to freeze, can, pickle and store for the winter months. If you do not want to bother with the perennials, reduce the size of your garden to approximately 30 feet wide by 50 feet long. If you are interested only in enough fresh vegetables to eat during the growing season, a 20- by by 30-foot garden is adequate.

Preparing the Soil

Do not work the soil when it is wet. To test for wetness take a handful of soil and squeeze it into a ball. If it does not readily crumble under slight pressure from the thumb, the soil is too wet to work. Also, if soil sticks tightly to garden tools, it is probably too wet.

To prepare a smooth, level, fine seedbed, rake or harrow the garden soil. Do not overwork soils with high silt and clay content. Leave them with the surface smooth but granular in texture rather than powdery fine.

If you have to turn under a heavy sod, do so as early in the spring as you can work the ground. This will allow time for the sod to partially decompose before you plant your crops.

Use a spade or plow, but turn the soil completely over. Spade or plow 6 to 8 inches deep. Spread manure, compost and fertilizer before spading. Rake or harrow at planting time, doing only that portion you intend to plant right away.

Lime

Most soils are naturally acid, but liming will correct this condition. A soil test will tell you how much lime you need to bring the garden soil to a pH of 6.5 (optimum for most garden vegetables). Upon request, you can obtain a soil-sample tube with information form and directions for use from the Agronomy Section, Plant Science Department, of your state university or

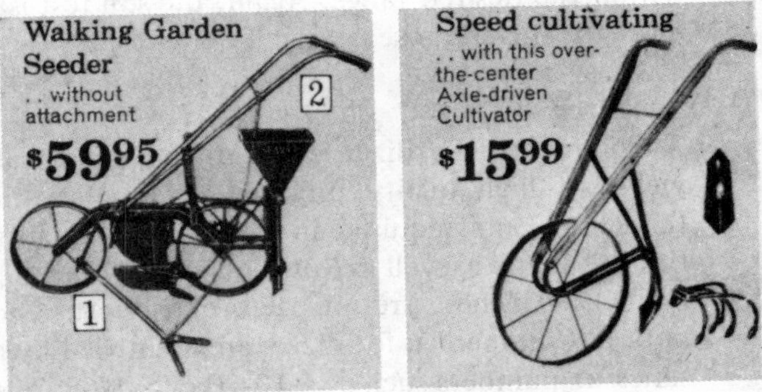

1 Seeder opens furrow, plants and covers seed, marks next row and all you do is walk. Sows every size garden seed. 7 planting speeds. Adjusts to any depth and spacing. 2-quart hopper.
32 KF 25962L3—Wt. 56 lbs... $59.95

2 **Fertilizer Attachment.** Sows fertilizer on both sides of seed as you plant .. saves time and effort. Hopper has 15-lb. capacity.
32 KF 25959C—Wt. 18 lbs.... $24.95

High-wheel cultivator aerates, mulches small or large areas effectively. 24-in. steel wheel provides easier rolling, maneuverability. Slotted foot changes cut or depth of angle according to work needs. Varnished oak handle provides sturdy frame, adjusts to desired height. Equipped with reversible steel shovel, 5-prong steel scratcher, mold board plow. Sent partly assembled by freight (rail, truck) or express.
71 KF 1843N—Wt. 21 lbs..... $15.99

Accessories and Implements for gas-powered roto-spaders

5 **Furrower.** Makes clean-cut planting or water furrows in spaded soil. Attaches to Roto-Spader stake bracket. Adjusts to various soil depths. Fits all Sears Roto-Spaders from 1962 to 1971.
32 KF 29015—Shipping weight 6 pounds.................... $7.49

6 **Dual-purpose Hiller.** Use as hiller with shovels spread apart .. move shovels together on crossbar and it becomes a furrower. Fits all Sears Roto-Spaders except pre-1968, 3 and 3½-HP models.
32 KF 29022C—Shipping weight 16 pounds................ $15.95

7 **Leveling Drag.** Levels out spaded soil, removes wheel and stake marks. Makes a fine seed bed .. eliminates the need for raking. Fastens to rear of all Roto-Spaders.
32KF29021–Shipping weight 11 pounds $13.95

8 **Til-row Attachment.** Makes it easy to put up your rows, cultivate your garden, make seed furrows, dig irrigation and drainage ditches. Fits all Sears Roto-Spaders.
32 KF 29068L—Shpg. wt. 20 lbs........ $34.95

9 **Steel Shields** prevent Roto-Spader from throwing dirt onto young plants; keep tines from tearing foliage in close work. Fits all Sears Spaders except pre-1968, 3 and 3½-HP models. Box of 2.
32KF29018–Shpg. wt. 6 lbs... Box $9.49

your county Extension Service office. You will receive precise recommendations for fertilizer and lime for your garden.

You will probably need to lime your vegetable garden every three to four years to maintain the desired pH. Do not overlime! Adding lime unnecessarily can affect the growth of your garden plants just as much as not adding it when needed.

Lime also provides plants with two nutrients necessary for plant growth—calcium and magnesium. For crops such as potatoes, where the recommended pH is only 5.0 to 5.5, keep a careful check on the availability of calcium and magnesium as nutrients even though the pH level is within the desired range. Again, the soil test is a *must* for this purpose.

Fertilizer

Commercial fertilizer is essential to maximum yields of high-quality produce. It supplies the three nutrients required in the greatest quantities by plants as well as nutrients usually lacking in the soil. These are nitrogen, phosphorus and potash. Commercial fertilizers are identified by a series of numbers such as 5-10-10 or 5-10-5. The first number from the left denotes the percentage of nitrogen in the fertilizer, the second number the percentage of phosphorus and the third number the percentage of potash. In general, figure on spreading 30 to 50 pounds of 5-10-10 fertilizer evenly over each 1,000 square feet of garden area. You can get a more precise measure of the fertilizer needed by having your soil tested.

If you make a second planting after the harvest of your spring-seeded crop, you will need less fertilizer. Plan on using about 2 pounds of fertilizer worked into the soil per 100 square feet.

Manure and Compost

Manure and compost add organic matter to the soil, which serves to improve aeration and water-holding capacity. It also tends to lessen compaction of finer-textured soils.

Manure is increasingly difficult to obtain. You can use horse or cow manure in the home garden in almost any quantity available. Besides its soil-improving qualities, it provides a portion of the nutrients required by the plants. When used as a fertilizer, add one pound of superphosphate per bushel of manure. One bushel of horse or cow manure would then be sufficient to fertilize about 50 square feet of garden area. But do not use poultry, sheep or goat manure at rates greater than one bushel per 100 square feet.

Compost provides a ready source of organic matter for the urban or suburban gardener. You can make a compost pile of almost any plant residue. Include leaves, trimmings, lawn clippings, corn husks and cobs, and any clean kitchen refuse such as peelings and cores. Try not to include crop residues from the vegetable garden, for fear of spreading plant diseases.

After the plant residues are assembled into a layer, cover it with a layer of soil or manure and add a small amount of fertilizer—about one pound to 25 to 50 pounds of organic matter. Then you may build additional layers upon this one. If the material used in the pile is dry, water each layer before you add the next one. Dish the finished pile on top to catch the water. In dry weather, it may be necessary to water the pile to get maximum decomposition. After several months, turn the pile over. The following season, when the pile has reached a suitable stage of decomposition, it will be nearly as rich as stable manure.

Planting

After carefully preparing your seedbed, measure and mark off the rows using a string as a guide. Open a trench with the corner of a garden hoe or a similar tool to the depth recommended in the planting guide. Place the seed in the bottom of the trench, cover with loose soil and firm gently.

When transplanting plants that you have grown or bought, be careful to keep as much soil around the roots as possible. The night before you intend to transplant, take an old knife and

Here is a plan for a large family garden with succession crops to increase yield, plenty of vegetables for winter processing, and widely spaced rows to allow for mechanical tillage.

Row				
1. Sweet corn	Early	Sweet corn	Midseason	
2. Sweet corn	Early	Sweet corn	Midseason	
3. Sweet corn	Early	Sweet corn	Midseason	
4. Sweet corn	Early	Sweet corn	Midseason	
5. Sweet corn	Late	Sweet corn	Late	
6. Sweet corn	Late	Sweet corn	Late	
7. Tomatoes (staked)				
8. Tomatoes (staked)				
9. Tomatoes (staked)				
10. Early Potatoes				
11. Early Potatoes				
12. Pepper		Eggplant		Chard (Swiss)
13. Lima beans (bush)				
14. Lima beans (bush)				
15. Lima beans (bush)				
16. Snap beans (bush)		This area should be replanted after the earlier harvest with endive, cauliflower, Brussels sprouts, spinach, kale, beets, cabbage, broccoli, turnips, lettuce, carrots and late fall potatoes		
17. Snap beans (bush)				
18. Broccoli				
19. Early cabbage				
20. Onion sets				
21. Onion sets				
22. Carrots				
23. Carrots				
24. Beets				
25. Beets				
26. Kale				
27. Spinach				
28. Peas				
29. Peas				
30. Lettuce	Seeded	Lettuce	2 weeks	Lettuce 4 weeks
31. Radish	Early	Radish	Later	Radish Later
32. Strawberries				
33. Strawberries				
34. Asparagus		Rhubarb		
35. Asparagus				

Feet between rows (top to bottom): 3, 3, 3, 3, 3, 3, 4, 4, 4, 3, 3, 3, 3, 3, 3, 3, 3, 3, 3, 3, 2, 2, 2, 2, 2, 2, 2, 2, 2, 2, 2, 2, 3, 3, 3, 3

N

100 feet

50 feet

cut between each plant so that each plant has its own block of soil. Make sure that the soil is moist, since this will help to hold it together during transplanting.

Dig a hole slightly larger than the ball of soil and set the plant a little deeper than it was in the flat or pot. Water each transplant with about a cup of starter solution or water. Starter solutions are soluble, high-analysis fertilizers and very useful in getting plants started more quickly. This quick start usually results in earlier maturity and increased yields. A number of commercial starter solutions are available at garden centers.

Weeding

Destroy weeds as soon as they appear. Once they become established they are hard to kill. They will use water and plant food needed by the vegetables.

One way to control weeds is through shallow cultivation. Scraping the surface with a hoe is sufficient if done when the weeds are small. Deep cultivation cuts off the roots of the vegetable plants and retards their growth.

Another method of controlling weeds is by the use of a mulch. Mulches also conserve soil moisture, reduce soil compaction and erosion and keep the vegetables clean.

Black polyethylene film is one of the more economical mulching materials. It lends itself particularly well to use with transplanted crops or larger seeded crops (like beans or corn) where it is laid before planting. Then punch holes and plant the crop through them. One disadvantage is that it must be taken up and thrown away at the end of the gardening season.

Organic residues form the other major source of mulching materials. You can successfully use straw, sawdust, wood chips, manure and many other materials. Organic mulches are normally applied after the crop has attained sufficient size so that it will not be buried by the mulch. For weed control with an organic mulch, apply it liberally. A dense material like sawdust or manure when applied should be at least 2 inches deep, while a loose mulch like straw should be a minimum of 6 inches deep. Because, as mulches decay they consume nitrogen, they will compete with the crop plants for plant food—unless you add more fertilizer. To compensate for this decay process, add an additional 3 pounds of 5-10-10 fertilizer to each 100 square feet of area mulched.

Watering

You must water heavily during dry periods. The needed frequency of watering will vary with soil texture, crop grown and many other factors. However, in most cases one thorough watering per week will be enough to give your garden plants maximum growth. A thorough soaking wets the soil to about 6 inches. To wet a sandy loam soil to this depth would require the application of about one inch of water while a clay loam would require 1½ inches. If you use sprinkler irrigation, you can measure the water applied by setting an empty can or other straight-sided container on the ground and catching the water in it.

Do not apply frequent light waterings since they encourage shallow rooting. Such shallow-rooted crops are then very susceptible to drought and more easily injured by cultivation.

Garden Pests

For a successful home garden, adopt some program of insect and disease control. You can maintain a good preventive program by the use of an effective general-purpose dust or spray. Such a material usually will contain one or two insecticides and a fungicide. The label will list the crops that can be treated and the insects and diseases which will be controlled. Read the label carefully and follow all precautions.

TREE FRUITS

Apples

If you plan to grow apples, you should certainly consider dwarf rather than standard trees. Here are some reasons:

Dwarf apple tree

1. Dwarf trees require less room.

2. They can be sprayed and pruned more easily.

3. The fruit can be picked without the use of long ladders.

4. More varieties can be grown in a limited space.

5. Dwarf trees come into bearing sooner.

What Stocks to Use for Dwarf Trees: The customary method of producing a particular type of tree is to bud or graft a section of wood of that variety onto a stock which grew from seed. Trees propagated in this way, unless restricted by pruning, may grow to 20 feet or more in height.

Other sources of stock on which to propagate a particular variety are known as E. Malling stocks. There are several of these, each of which is designated by a Roman numeral ranging from I to XVI. By a special method of propagation,

all stocks within a particular group, such as E. Malling I, are uniform in all essential characteristics and have the same effect on the size of the trees propagated from them. Thus trees propagated on E. Malling I will be smaller than those propagated on E. Malling XVI, which will grow to full size, the extent of dwarfing depending upon the particular E. Malling stock used.

Trees propagated on E. Malling IX will be true dwarf, growing to a height of only 10 to 12 feet at maturity. This is the type usually recommended for planting in the home garden, since they require the least room and start to bear the second or third year. When mature, they will produce one or more bushels of apples fully as large as those produced on standard trees.

Trees propagated on E. Malling IX are shallow rooted and therefore need support to keep them upright. It is common practice to tie the trees to wooden stakes or to pieces of pipe driven down beside them.

Many nurseries advertise dwarf trees of the more common varieties. When purchasing such trees you should specify those propagated on E. Malling IX.

Pears

Pear trees make an excellent addition to the home garden and require less spraying than apples. Pear buds and wood are hardier to winter cold than peaches, but not as hardy as some apple varieties. Depending on the variety grown, the harvest season of pears is from mid-August to October in Massachusetts.

It is possible to obtain a limited number of pear varieties as dwarf trees. Pears are dwarfed by propagation on quince root stock. However, not all pear varieties grow satisfactorily on quince roots. This can be overcome by first budding a compatible variety on the quince root, then following a year's growth, budding the desired variety on the new stem. When buying dwarf pear trees, request those on which the double-budding technique has been used.

Peaches

Picking a luscious tree-ripened peach from your own tree is one of the most satisfying

PEARS

HEAVY BEARING MONEY-MAKERS

They'll Melt in Your Mouth

BOSC-November, December. Russet. Extra large.
BARTLETT-September. Probably best pear for all use.
CLAPP'S FAVORITE-August. Fruit, juicy, fine grained.
DUCHESS-Autumn. A good canning pear. Large juicy.
FLEMISH BEAUTY-September, Large sweet pear
SECKEL-September 15th. Used for pickling.
KEIFFER-Good to eat at Thanksgiving.

PRICE ON STANDARD PEAR TREES

Your choice of varieties, alike or assorted.

Size		Each	3-9 Each	10-49 Each	50 up Each
11/16	5-6 ft....	$3.25	$2.95	$2.45	$1.95
9/16	4-5 ft....	2.95	2.65	2.25	1.70
7/16	3-4 ft....	2.65	2.40	2.05	1.55

DWARF PEACH TREES

Striking beauty, early bearing lucious full size fruit. Full maturity at only 6 to 7 feet.
VARIETIES: Elberta, Red Haven, Golden Jubilee
PRICE: Each..$4.25; Three or more,Each...$3.95

Dwarf Pear Trees

Produce Large, Full Size Pears Will Take Only a Corner of Your Yard

These extra-hardy Dwarf Pears thrive almost anywhere. Heavy bearing. Bloom in spring. We sell strong, well-rooted trees.

VARIETIES: Bartlett, Duchess, Seckel.
Your Choice of Varieties, Alike or Assorted

Size	Each	3-9 Each	10-49 Each	50-up Each
2 year, 3-4 ft.....	$4.25	$3.95	$3.70	$3.20
2 year, 2 1/2-3 ft..	3.95	3.70	3.45	2.95

Dwarf Plum Trees

Plant trees 8 to 9 feet apart. Self-pollenizing, but will produce heavier crops if two trees are planted.
STANLEY-Newest and most popular. Large, freestone, dark blue Prune Plum
BURBANK-An old favorite; bright reddish purple Delicious quality and prolific bearers.
2 year trees, Each....$4.25; Two for......$7.95

Dwarf Nectarines

Enjoy the superb quality found only in nectarines by planting these fine dwarf trees this season.
VARIETIES: Garden State and Redchief.
PRICE: $4.25 each: Two for..............$7.95

Dwarf Apricots

Grow about 6 to 8 feet tall. Easy to harvest. Do well wherever peaches are grown.
VARIETIES: Early Golden and Moorpark.
EACH: 3-4 ft....$4.50 2 for...........$8.25.

experiences of home gardening. But a peach crop is less certain than most tree fruits and you must be prepared for disappointment.

Peach trees are subject to injury from low temperatures. The buds which produce flowers in the spring are formed during the previous summer, and the potential crop is present on the trees during the winter. The critical temperature at which peach buds will be killed is about −15° F. If peach trees are planted in areas where the winter temperature drops to this point or lower, there may be a total loss of the crop the following summer. If the winter temperature goes to −20° F. or lower, not only will the buds be injured but the trees themselves may be killed.

It is unsafe to grow peaches if chokecherries are growing in the vicinity, since these harbor the virus disease red-yellow virosis, or X-disease, which cannot be controlled by spraying. Unless the area within 500 feet of the prospective planting is free from chokecherries, you should not undertake peach growing. Information on identification and methods of eradicating chokecherries is contained in Special Circular 216, "Chokecherries" (New York State College of Agriculture, Cornell University, Ithaca, N.Y. 14850).

Since peach trees are normally smaller in stature than standard apple and pear trees, dwarf trees may not be necessary for the home garden. However, if space is definitely limited, dwarf peach trees may be desirable and are available from some nurseries.

Plums

While the fruits mentioned so far are generally the most popular of the tree fruits, you should not ignore the possibility of adding a few plum trees to your garden. Trees of the recommended plum varieties are hardy in the northeast and will provide delicious fruit from mid-August until the middle of October.

There are two types of plums desirable for the home garden: European and Japanese. They differ in habit of tree growth and general characteristics of the fruit. The trees of the European type are typically upright, while the Japanese are spreading or drooping. Most European plum varieties are blue or yellowish-green in color, while most Japanese varieties are red.

Where space is sufficient to permit a planting distance of about 20 feet each way, standard trees are to be preferred, but if this amount of space is not available, dwarf plum trees planted 12 feet apart are the answer.

Cherries

Before attempting to grow any type of cherry, be aware that birds will harvest most or all of the crop unless provision is made to protect the fruit from them. The most practical means is to cover the trees with cheesecloth or tobacco cloth as the fruit starts to ripen. Also cherries, like other stone fruits, are susceptible to brown rot and the crop may be a complete failure unless this disease is controlled.

While sour cherries are relatively hardy in low winter temperatures, sweet cherries may be severely injured, if not killed. However, your county farm agent can tell you how sweet cherries can be grown successfully by using the sod culture method with a limited amount of nitrogenous fertilizer.

Since most varieties of sweet cherries make for large trees, dwarf trees are a good idea for the home garden. Sour cherry trees, on the other hand, are normally dwarfish so there is little advantage to planting dwarfs in this case.

Choice of Trees

You should purchase the fruit trees from a reliable nursery or garden-supply center. It makes little difference where the trees are grown, except that trees grown at distant locations have more chance of drying out in shipment than if grown nearby.

In general, one-year-old trees are preferred to older ones, as less of the root system is lost in proportion to the top when the trees are dug. Two-year-old trees are of course larger than one-year trees and have several branches, whereas one-year trees usually consist of a single unbranched shoot. The latter are frequently referred to as "one-year-whips." One-year peaches, sour cherry and some plum varieties, however, are usually branched, while one-year apple, pear and sweet cherry are typically single shoots.

Whether you buy dwarf or standard trees, unless they are sour cherry or peach trees, it is essential that you plant at least *two varieties* of a particular kind of fruit in order to provide for cross-pollination.

Preparation of the Soil

Fruit trees should be planted as early in the spring as the soil can be easily worked. The soil may be plowed and disked prior to planting. But it is possible, by application of sufficient mulching materials around the trees, to maintain satisfactory growth on fruit trees set in heavy grass sod.

Dwarf trees will not thrive in competition with weeds or tall grass. With adequate fertilization, they may be grown successfully in a clipped lawn, in sod with additional mulch applied or under cultivation.

Fruit trees grow best in a soil that is slightly acid (pH 6.0 to 6.5). Have your soil tested to determine pH and apply sufficient high-magnesium lime to maintain the 6.0 to 6.5 pH. Lime may be applied anytime during the year.

Planting the Trees

Dig the hole for the tree large enough to take in the whole root system. Set the tree in the soil about the same depth as it was grown in the nursery. Plant dwarf trees with the graft union above the ground. The graft union is indicated by a slight swelling, a change in bark color or a slight bend in the trunk near the root.

In planting a tree the soil should be thoroughly tamped around the roots so that they will be in contact with moist soil. It is not necessary to add water to the soil unless it is dry. Put no fertilizer or manure in the hole at planting time, as there is too much risk of injury to the tree roots from the concentrated fertilizer solution.

Pruning

Training and pruning young trees is important in determining the strength and bearing capacity of the tree throughout its productive life. If you fail to prune a young fruit tree carefully, the age at which it starts to bear a profitable crop will be delayed and the tree may develop weak crotches which will break down under the first heavy crop.

Rodent and Rabbit Control

Mice are a serious pest in young orchards and can kill or permanently injure trees. The use of wire hardware cloth guards around the trunk of the tree and several inches of sand applied to the base and extending in a 2-foot circle help reduce mouse injury. The hardware cloth must have three or four wires to the inch to be mouseproof. Push the base of the hardware cloth into the sand to help prevent mice tunneling under the guards and girdling the trees. Keeping the soil area around the tree trunk free of tall grass and weeds during the late summer and fall is of special value in preventing girdling.

Rabbits and woodchucks sometimes cause considerable injury to young trees, and wire guards help to prevent such damage. The rabbit population can be reduced by hunting and trapping, while woodchucks can be controlled by placing gas bombs in their burrows. Also, application of repellents for the protection of fruit trees is effective in preventing rabbit damage. The gas bombs and directions for their use as well as information on repellents may be obtained from the U.S. Fish and Wildlife Service, Washington, D.C.

14

Dairy Products

MAKING COTTAGE CHEESE AT HOME

Cottage cheese is tasty, nutritious, easily digested, and surprisingly low in calories. Eat it plain; add a little salt and pepper or a dash of some other seasoning; combine it with fruits or vegetables in a refreshing salad that's a main dish or a dessert. Keep a supply in your refrigerator or springhouse for a snack. If you don't have a springhouse, this is as good a place as any to show you how to build one. (See next page.)

Cottage cheese—a soft, unripened cheese—can be easily made at home from skim milk. The freshly made curd has a mild acid flavor and smooth texture. It contains about twenty percent milk solids, and many of the same nutrients found in fresh milk.

Most homemakers who make their own cottage cheese like having a steady supply of cheese with homemade flavor. Families that have large quantities of surplus skim milk for making into cheese may save money, too.

TYPES OF COTTAGE CHEESE

With and Without Rennet

The two major types of cottage cheese are small-curd, high-acid cheese made without rennet, and popular large-curd, low-acid cheese made with rennet.

Rennet is a substance that speeds curdling and keeps the curd that forms from breaking up easily. Adding rennet shortens the cheese-making process, results in a less-acid and larger-curd cheese and reduces the amount of curd poured off with leftover liquid.

With Cream

Cottage cheese made either with or without rennet can be creamed. Adding cream to cheese increases its smoothness and improves its flavor and texture. Creaming cottage cheese also adds calories and slightly lowers protein content.

Fruits, vegetables or other flavorful foods are often added to cottage cheese to make a variety of side dishes and salads.

INGREDIENTS

Milk

Use pasteurized skim milk or nonpasteurized milk. One gallon (8.25 pounds) of milk will yield about one pound of cottage cheese. The equipment that is specified in this chapter will

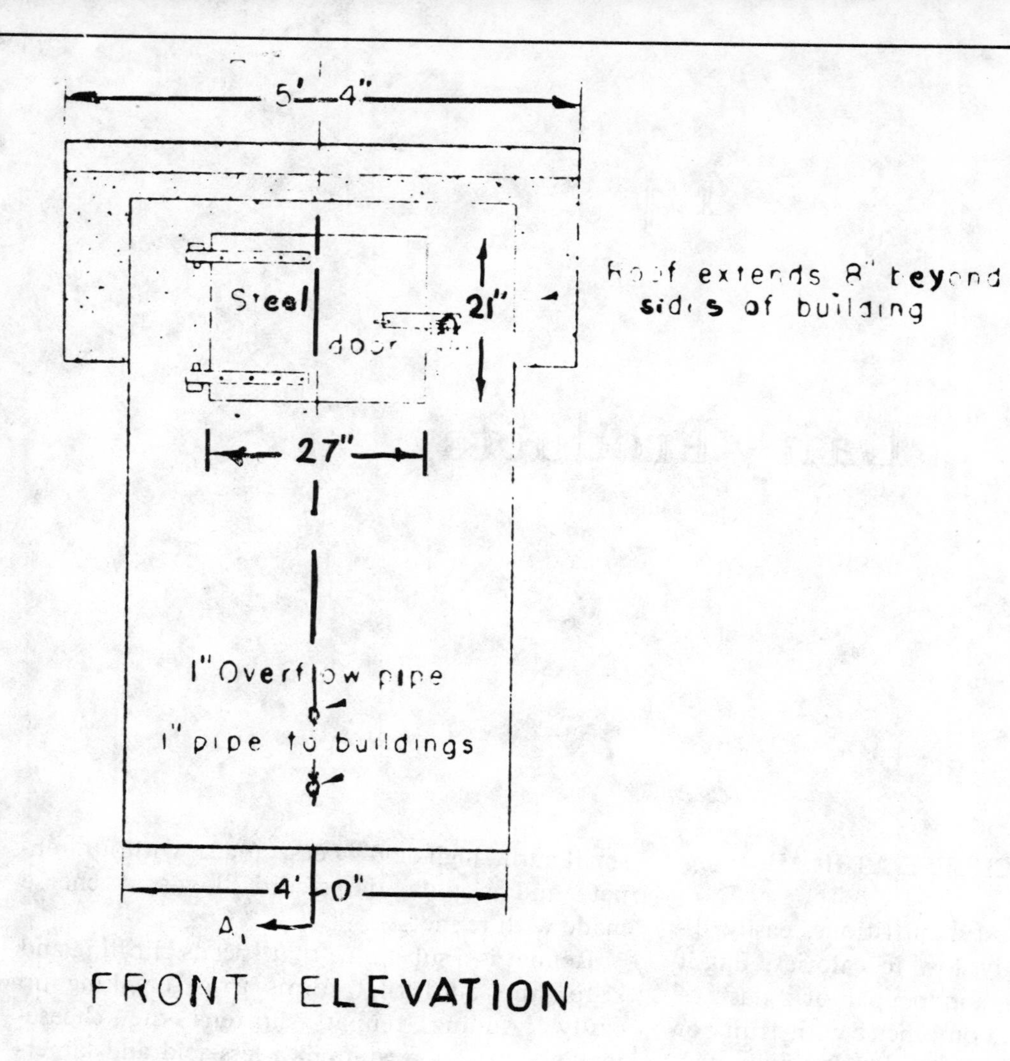

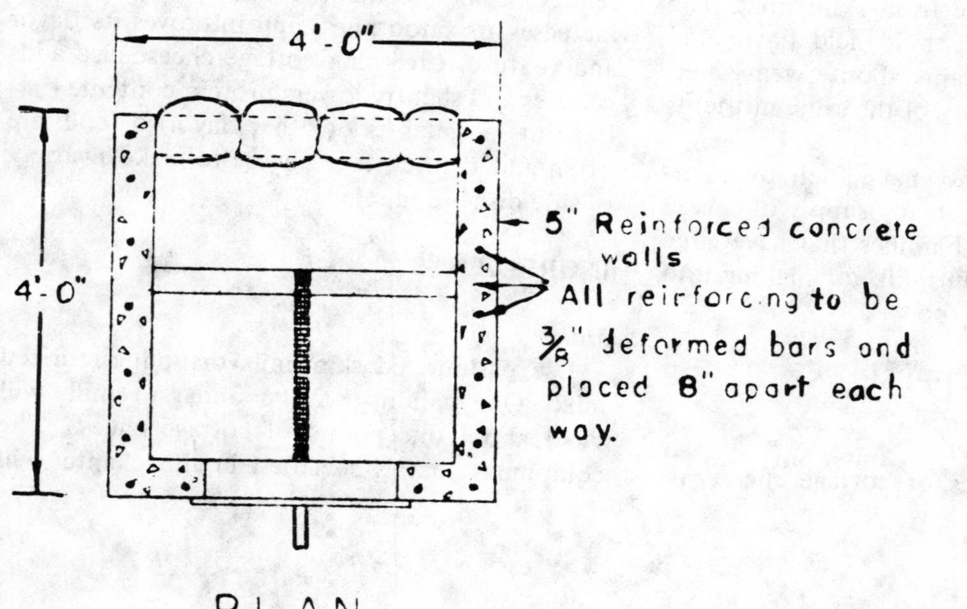

Plans for construction of typical springhouse found on many farms in the United States and Canada. It is built over ice-cold spring bubbling from ground. Its insulated walls and roof maintain an interior temperature of 45° F. to 50° F. the year round, furnishing farmwives a natural cooling room for storing dairy products and other foods without the need of refrigeration.

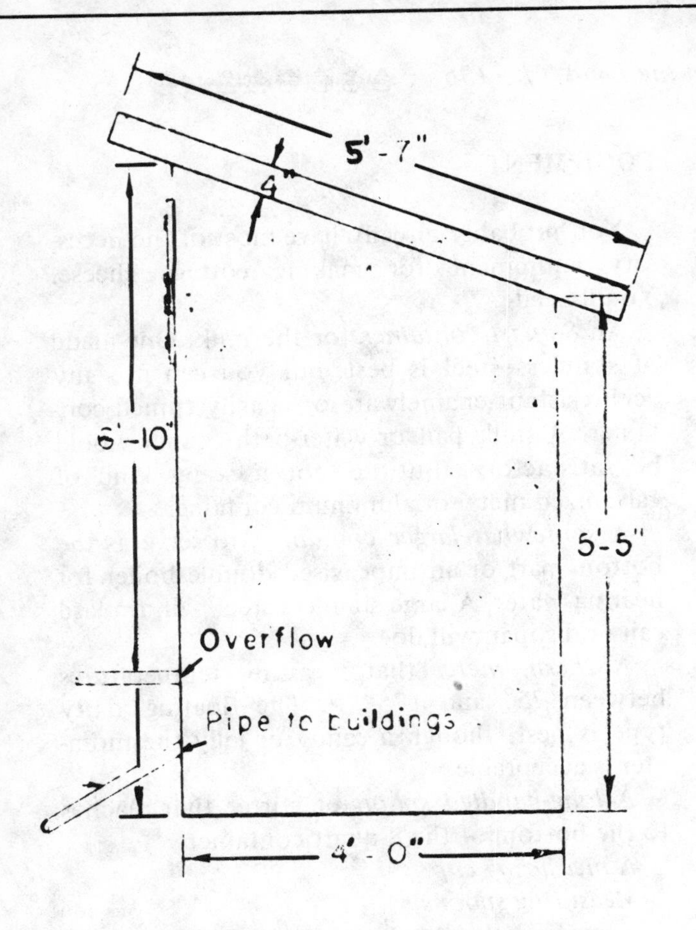

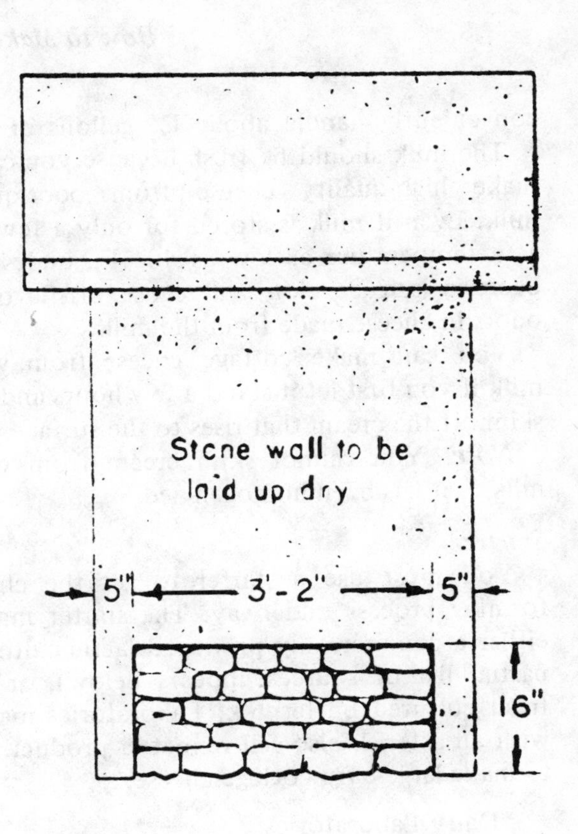

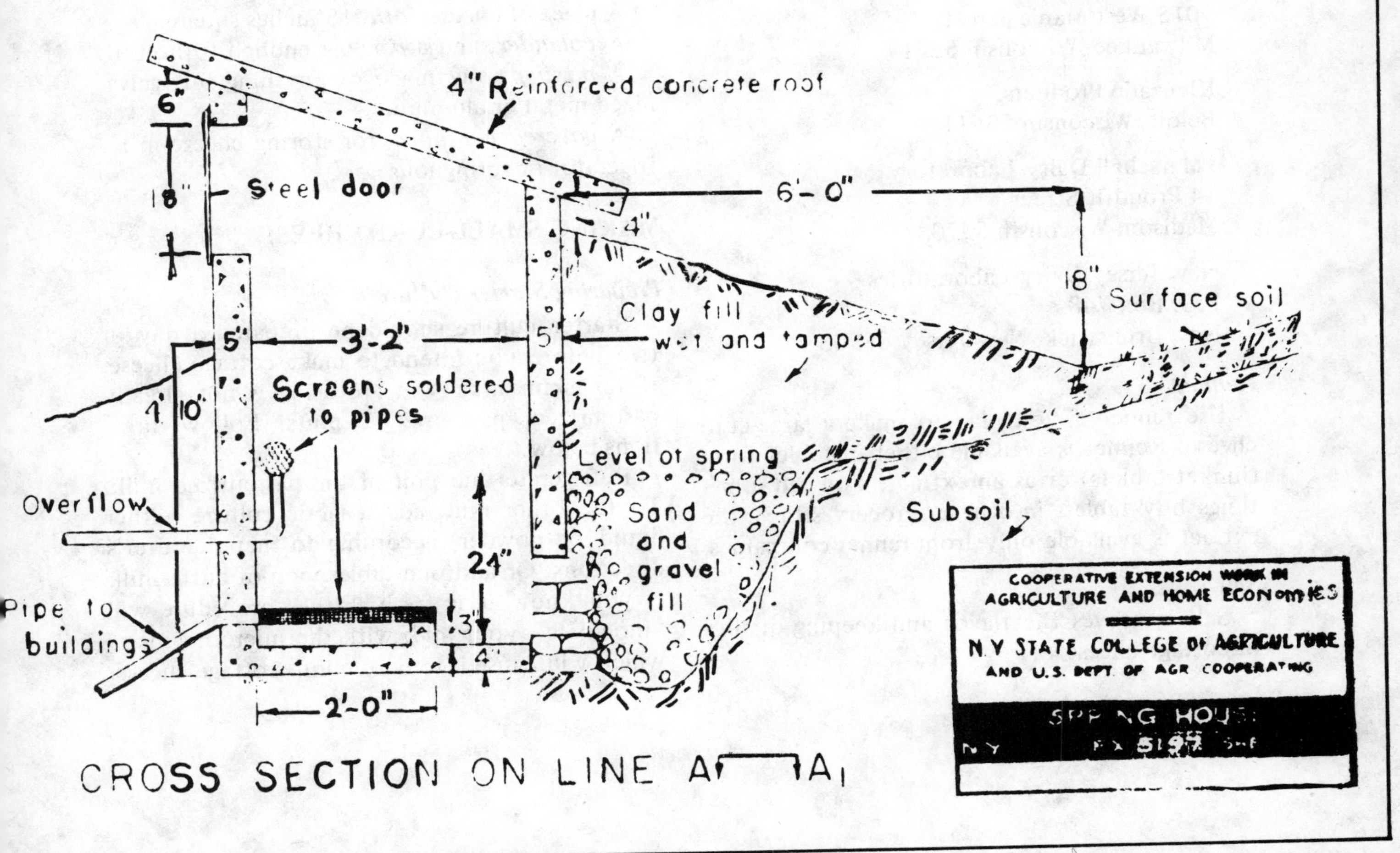

COOPERATIVE EXTENSION WORK IN
AGRICULTURE AND HOME ECONOMICS
N.Y. STATE COLLEGE OF AGRICULTURE
AND U.S. DEPT. OF AGR. COOPERATING
SPRING HOUSE

conveniently handle about 1½ gallons of milk.

The milk should be fresh because you cannot make high-quality cheese from poor-quality milk. Even if milk is stored for only a few days at a temperature as low as 40° F., undesirable bacteria can develop and cause off-flavors or odors in cheese made from the milk.

You can make cottage cheese from whole milk if you first let it stand a few hours and then skim off the cream that rises to the surface.

Note: You cannot skim cream from whole milk that has been homogenized.

Starter

You must use a starter to get the cheese-forming process underway. The starter may be either a commercially produced lactic culture (a partial list of sources appears below), or very fresh cultured buttermilk. Laboratories marked with an asterisk also sell rennet, a product used to make large-curd cottage cheese.

*Dairy Laboratories
2300 Locust Street
Philadelphia, Pennsylvania 19103

Chr. Hansen's Laboratory
9015 West Maple Street
Milwaukee, Wisconsin 53214

Klenzade Products
Beloit, Wisconsin 53511

*Marschall Dairy Laboratory
14 Proudfit Street
Madison, Wisconsin 53703

New Jersey Dairy Laboratories
P.O. Box 748
New Brunswick, New Jersey 08903

Rennet

Use rennet if you plan to make a large-curd cheese. Rennet is available either in tablet form (junket tablets) or as an extract. You can sometimes buy tablets in drug or grocery stores; the extract is available only from rennet companies.

Salt

Salt improves the flavor and keeping quality of cottage cheese.

EQUIPMENT

You probably already have most of the necessary equipment for making cottage cheese. You'll need:

An *8-quart container* for the milk. One made of stainless steel is best, but you can use any acid-resistant enamelware or heavily tinned container. A milk pail or water-bath canner would be satisfactory. But do not use any kind of galvanized metal or aluminum container.

A *somewhat larger container*, to serve as the bottom part of an improvised double boiler for heating water. A large stainless steel pail, tin lard can or dishpan will do.

A *thermometer* that measures temperatures between 75° and 175° F. The floating, dairy type is best, though a candy or jelly thermometer is acceptable.

A *long-handled spoon* or stirrer that reaches to the bottom of the 8-quart container.

A *measuring cup.*

Measuring spoons.

A *knife* with a blade long enough to reach to the bottom of the 8-quart container.

A piece of *cheesecloth*, 18 inches square.

A *colander*, and a *pan* big enough to hold it.

A *mixing bowl* made of anything but galvanized metal or aluminum.

A *covered container* for storing cheese in refrigerator or springhouse.

MAKING SMALL-CURD CHEESE

Preparing Starter Culture

Starter culture should be prepared a day or two before you intend to make cottage cheese. If the skim milk you are using is not already pasteurized, pasteurize 2 pints. Follow directions below.

Refrigerate one pint of the pasteurized milk. To the other pint, add a lactic culture (either liquid or powder) according to manufacturer's directions. Or add one tablespoon of buttermilk if you know it is fresh. Either procedure will "inoculate" your milk with the microorganisms which will cause it to curdle into cottage cheese.

Hold the inoculated milk at 70° to 75° F. for 16 to 24 hours, or until it curdles.

With a scalded and cooled teaspoon, add a teaspoon of the curdled milk to the pint of milk you have been keeping in the refrigerator.

When this second culture has also curdled (in 12 to 18 hours), it is ready for use as a starter to ripen milk for making into cottage cheese.

How to Pasteurize Milk

All the milk you use in making cottage cheese should be pasteurized. Pasteurization will kill harmful bacteria and most of the organisms that may produce off-flavors in cottage cheese.

Almost all fluid skim milk that you buy has already been pasteurized, so you won't have to pasteurize it yourself. However, if you are using nonpasteurized milk from your own cows, pasteurize it first.

If you have an electric, commercially made pasteurizer, simply follow the manufacturer's directions. Otherwise, improvise a large double boiler and follow this method:

Heat water in outer container until the temperature of the milk in the inner container reaches 145° F.

Keep milk at this temperature for 30 minutes. (You will probably have to adjust the heat to maintain the temperature of the milk at 145° F. throughout the half hour.)

Cool the milk to 72° F. You can do this by simply emptying the outer container and refilling it with cold water.

Start making cottage cheese immediately or cool the milk to at least 50° and refrigerate or store in springhouse for later use.

Warming the Milk

Milk to be made into cottage cheese should be at room temperature (about 72° F.). You should maintain this temperature throughout the cheese-forming process until curd is formed, cut and ready for final heating.

Warm the milk by placing it and its container inside a larger container filled with water. Heat the water until milk reaches room temperature (72° F.).

Curdling the Milk

Do not stir the milk. Let it stand at room temperature for 16 to 24 hours. (Occasionally you may have to reheat the water in the outer container, to maintain the temperature at 72° F. in the inner one.)

For the best cheese, your milk should curdle over the 16- to 24-hour standing period. If it curdles sooner, use less starter the next time you make cheese. If it does not curdle satisfactorily during this time, use more starter next time.

When curdling occurs, a jellylike, firm substance (curd) forms and a small amount of watery liquid (whey) usually appears on the surface. To determine if the curd is ready for cutting, insert a knife or spatula into the curd at the side of the container and gently pull the curd away from the container side. If the curd breaks quickly and smoothly, it is ready to be cut.

Cutting the Curd

Cutting the curd into ¼-inch pieces requires the four steps described:

Insert knife blade through the curd to the bottom of the container on the side opposite you. Then pull the knife, held vertically, toward you. Withdraw the knife and repeat the cutting, every ¼ inch.

Turn the container a quarter-turn. Repeat the first step, again cutting the curd every ¼ inch.

Turn the container to its original cut line and cut the curd at an angle.

Again turn the container and repeat the cutting.

When the curd is cut roughly into ¼-inch pieces, let it stand for 10 minutes. During this time, whey separates from the curd and the curd begins to become slightly firmer (although it is still much too soft to be stirred).

Heating the Cut Curd

This is a critical step in making cottage cheese.

Add water (72° F.) to the outer container until its level is slightly above the level of curd and whey in the inner container.

Heat the water slowly and as uniformly as possible, to raise the temperature of the curd and whey to 100° F. in 30 to 40 minutes—a temperature increase of about one degree per minute.

During heating, stir the curd gently with a large spoon—about a minute at a time, every 4 or 5 minutes. This helps heat the curd uniformly and prevents curd particles from sticking together.

When the curd and whey reach a temperature of 100° F., heat faster and stir more frequently. The temperature of the curd and whey should reach 115° in 10 to 15 minutes. Hold at this temperature for 20 to 30 minutes or until the pieces are firm and do not break easily when squeezed.

If the curd doesn't become firm enough at this temperature, heat it to 120° F., or even to 125°.

Stir the curd and whey constantly and test the curd often for firmness. When the curd is firm enough, stop the heating process.

Removing the Whey

When the curd has firmed sufficiently, dip off most of the whey.

Pour the remaining curd and whey into a fine-meshed cheesecloth spread over a colander that you've set in the sink or in another pan. Let the curd drain for 2 or 3 minutes.

Note: Don't let the curd drain too long or curd particles will stick together in large clumps.

Washing and Cooling the Curd

Gather together the corners of the cheesecloth containing the curd. Immerse both cloth and curd in a pan of clean, cool water. Raise and lower the "bag" of curd several times, for 2 or 3 minutes, to rinse the whey from the curd and to cool the curd.

Rinse the curd again in ice water, for 3 to 5 minutes, to chill it.

Put the curd in a colander set inside a larger pan. Shake the colander occasionally, until the whey stops draining.

If you prefer unsalted, uncreamed cottage cheese, you can now remove the curd from the cloth, pack it in a suitable container and store it in the refrigerator or springhouse. However, unsalted cheese will have a definite acid taste.

Salting the Curd

After transferring the curd from the cheesecloth to a mixing bowl, add a teaspoon of salt for each pound of curd. Mix thoroughly.

Creaming the Curd

For each pound of curd, add 2 or 3 ounces (4 to 6 tablespoons) of either sweet or sour cream, or of half-and-half. Mix thoroughly.

REASONS FOR IMPERFECT COTTAGE CHEESE

Sour acid flavor means that too much acid developed before and during cooking of the curd, that too much whey was retained in the curd or that the curd was not sufficiently washed and drained.

Yeasty, sweet or unclean flavors indicate that yeasts, molds or bacteria were introduced into your cheese by unclean utensils or an impure starter; or that your milk was not completely pasteurized.

Soft wet curd results from the development of too much acid during cutting of the curd, heating the cut curd at too high or too low a temperature or allowing too-large curd particles to form.

Tough dry curd results from insufficient acid development in the curd before it is cut, too fine a cutting of the curd, too high a heating temperature or too long a holding time after cooking and before dipping off the whey.

MILK

Milk has a well-balanced protein, carbohydrate, fat and mineral content, and it contains most of the essential vitamins.

Milk should be cooled immediately after it

comes from the cow; keep it closely covered in sterile containers, preferably of glass, earthenware or enamel. Uncovered milk absorbs odors and collects bacteria, especially when warm. Milk is preserved by pasteurization and, as already directed, may be pasteurized at home.

The vitamin D content of milk can be increased by feeding vitamin D to the cow, by irradiating the milk directly or by adding a vitamin D concentrate.

Certified milk is not pasteurized. It is milk produced and bottled under the sanitary regulations laid down by your state Milk Commission.

Unsweetened evaporated milk is simply pure whole milk; the water is removed and nothing is added. It *is* homogenized and may or may not be enriched with "sunshine" vitamin D. Evaporated milk mixed with an equal amount of water can be used in any recipe calling for whole milk.

Sweetened condensed milk is also pure whole milk; the water has been extracted, and sugar added.

Cream is fat that rises to the top of nonhomogenized milk if left standing. Skim milk is produced by skimming off the cream. When the cream is churned into butter, the liquid that remains is buttermilk.

To produce sour milk for cooking, keep milk undisturbed in a shallow covered pan at a temperature of 90° to 100° F. until it becomes thick and clabbered. Should it sour too slowly, it will become bitter. To substitute sour milk for sweet milk in baking, add ¼ to ½ teaspoon of baking soda to one cup of sour milk for every 2 cups of flour.

Another way to sour milk is to add one tablespoon of lemon juice to one cup of sweet milk, or to use one tablespoon of vinegar instead of the lemon.

HOMEMADE SWEET BUTTER

Both my grandmother and mother hand-churned their own sweet butter on the farm. It was a long time ago, but I still remember the way home-baked bread, biscuits, cornbread and rolls tasted when spread with farm-churned butter, which melted into golden pools on the freshly baked, heavenly smelling delights still hot from the oven.

When sweet cream is readily available, homemade sweet butter on the table more often then not separates the good cook from the bad.

Years ago it was the children or the husband who counted the strokes and took turns lifting up the dasher of the butter churn through the cream and pushing it down again. Today hand churns are still used, but an electric mixer does the work faster and more easily.

If you own cows and have your own separator, use one to 1½ quarts of heavy cream from the separator (or buy the cream at your dairy store).

Cream from your own cows should be pasteurized in an electric pasteurizer or heated well over a low flame until it begins to rise in the pan. Remove from heat and let the cream settle down. Repeat this same process two more times.

A glass hand churn with wooden paddles for making small quantities of butter

Cool by setting the pan in cold water. Then pour into sterilized glass jars, cover and store for several days in a refrigerator or springhouse. Cream kept cold for two to four days will churn much faster than if 24 or less hours old.

After several days' storage, "ripen" the cream by letting it stand at room temperature for at least 6 hours. It will thicken and become mildly sour. This gives the butter a mild and wonderful flavor. Once the cream has ripened, it should be cooled again before churning can begin.

Pour the cooled cream into the largest bowl of your electric mixer (or into a hand churn or an electric butter churn). If a mixer or an electric butter churn is used, beat at high speed until flecks of butter begin to form and float to the surface. Then turn to low speed until butter separates from milk. As the butter forms, it will cling to the sides of the bowl or churn. Keep pushing it down the sides with a spatula.

Now pour off the buttermilk, which is the liquid left after formation of the butter, measure it out in a measuring cup and add the same measure of cold water to the bowl. Still beating at low speed, continue pouring off the water and replacing it with fresh water until the butter clots and thickens.

Remove beaters, scrape off the butter with a spatula and use it to work out water by pressing the butter against the sides of the bowl. Be sure to press out *all* the water. Now mold the butter in a butter press or empty it into a container with a tightly fitting lid. If it is to be kept for any length of time before using, store it in a refrigerator, freezer or other cold facility.

Homemade butter has a very light color. Butter purchased at the store has been dyed yellow by food colorings.

15

Farm Canning of Fruits and Vegetables, and Home Storage

Organisms that cause food spoilage—molds, yeasts and bacteria—are always present in the air, water and soil. Enzymes that may cause undesirable changes in flavor, color and texture are present in raw fruits and vegetables.

When you can fruits and vegetables you heat them hot enough and long enough to destroy spoilage organisms. This heating (or processing) also stops the action of enzymes. Processing is done in either a boiling-water-bath canner or a steam-pressure canner. The kind of canner used depends on the kind of food being canned.

For *fruits, tomatoes and pickled vegetables,* use a boiling-water-bath canner. You can process these acid foods safely in boiling water.

For *all common vegetables except tomatoes*, use a steam-pressure canner. To process these low-acid foods safely in a reasonable length of time takes a temperature higher than that of boiling water.

A pressure saucepan equipped with an accurate indicator or gauge for controlling pressure at 10 pounds (240° F.) may be used as a steam-pressure canner for vegetables in pint jars or No. 2 tin cans. If you use a pressure saucepan, add 20 minutes to the processing times given for each vegetable.

Steam-Pressure Canner

For safe operation of the canner, clean the pet cock and safety-valve openings by drawing a string or narrow strip of cloth through them. Do this at the beginning of the canning season and often during the season.

Check the pressure gauge. An accurate pressure gauge is essential.

A weighted gauge needs to be thoroughly clean.

A dial gauge should be checked *before* the canning season.

If your gauge is off 5 pounds or more, buy a new one. If the guage is not more than 4 pounds off, you can correct for it as shown below.

The food is to be processed at 10 pounds steam pressure.

SEARS CANNING EQUIPMENT

Brighten, enrich winter meals and save time, money in meal preparation
Can stocks and soups • homemade relishes • wild game • seafood • fruits and vegetables

1 Aluminum Blancher-Cooker. Large 8-quart cooker with 7-quart insert.
Shipping weight 2 lbs. 8 oz.
11 AF 1266 . . . . $5.99

2 Steel Blancher-Cooker. Lustrous porcelain enameled finish. 7-quart liquid capacity. 5½-quart perforated insert.
Shipping weight 3 pounds.
11 AF 5125 . . . . $3.99

3 Bottle Capper. Metal. Caps 8½ to 12-inch tall bottles easily . . quickly.
Shipping weight 5 pounds.
11 AF 5118 . . . . $7.49

4 Lacquered Bottle Caps. Metal with air-tight cork lining . . keeps flavor in, odors out. Box of 144. Caps in assorted sizes.
Shipping weight 1 pound.
11 AF 5117 . . . . Box 99c

5 Apple Parer. Also slices, cores. Metal frame. Shpg. wt. 2 lbs.
11 AF 5116 . . . . $5.49

6 Tin Can Sealer. Fits numbers 2 and 3 cans. Metal. Can not included. Shipping weight 11 lbs.
11 AF 5147 . . . . $17.99

7 Tin Cans with lids. Plain or marked with "R" for red-meated items. Not reusable.

No. 2 Plain Can. Box of 100.
Shipping weight 25 pounds.
11 AF 5159L . . . . Box $12.99

No. 3 Plain Can. Box of 100.
Shipping weight 35 pounds.
11 AF 5160N . . . . Box $18.99

No. 3 "R" Cans. Box of 100.
Shipping weight 35 pounds.
11 AF 5157N . . . . Box $19.99

Shipping Note. 11 AF 5160N and 11AF5157N shipped freight (rail or truck) or express.

Economical *Water-bath Canners* . . fine for fruit, tomatoes and pickled foods

8 Steel Canner. Porcelain enamel finish. Wire jar rack. Jars not included.

9-jar. 35-quart capacity. Shipping weight 8 pounds.
11 AF 5119L $6.49

8-jar. 24-quart capacity. Shipping weight 6 pounds.
11 AF 5120 . . . . $4.99

9 Aluminum Canner. 20-quart liquid capacity. Rust-resistant steel rack holds 7 jars (jars not incl.). Service weight. Polished outside . . natural finish inside. Shpg. wt. 5 lbs.
11 AF 5105 . . . . $5.99

Speedy Pressure Cooker-Canners . . a bacteria-safe method for meats, vegetables—and it takes less than half the time

The **all-foods** processing method: meats of all kinds and all vegetables except tomatoes, sauerkraut, and ripe pimientos are low-acid foods with little natural defense against bacteria. They must be processed in pressure cooker which reaches many degrees above the boiling point of water.

(10 and 11) Heavy gauge cast aluminum. Resists rust. Pressure regulator to vent steam. Recipes, instr.

10 16-quart Canner. Basket holds 9 1-pt., or 7 1-qt. jars; 13 No. 2 or 5 No. 3 tin cans (not incl.).
11 AF 4632C—Shipping weight 16 lbs. $21.99

11 Canners with pressure gauges. No guessing about vital timing, you see when pressure is right.

16-qt. Basket holds 9 1-pt., 7 1-qt. jars; 13 No. 2, 5 No. 3 cans (not incl.).
Shipping weight 17 lbs.
11 AF 4600C $29.99

21-qt. Basket holds 18 1-pt., 7 1-qt. jars; 17 No. 2, 10 No. 3 cans (not incl.). Shpg. wt. 18 lbs.
11 AF 4601L $32.99

If the gauge reads high:

1 pound high—process at 11 pounds.
2 pounds high—process at 12 pounds.
3 pounds high—process at 13 pounds.
4 pounds high—process at 14 pounds.

If the gauge reads low:

1 pound low—process at 9 pounds.
2 pounds low—process at 8 pounds.
3 pounds low—process at 7 pounds.
4 pounds low—process at 6 pounds.

Water-Bath Canner

Any big, metal container will do for a boiling-water-bath canner; it must be deep enough to have an inch or two of water over the tops of the jars and a little extra space for boiling. Use a cover under the rack to keep the jars from touching bottom.

The rack may be of wire or wood. Have partitions in the rack to keep the jars from touching one another or falling against the side of the canner.

If a steam-pressure canner is deep enough, you can use it as a water bath. Set the cover in place without fastening it. Be sure you have the pet cock wide open, so that steam can escape and no pressure is built up.

Glass Jars

Jars and lids must be perfect. Discard those with cracks, chips or dents.

Wash glass jars in hot, soapy water and rinse well. Wash and rinse all lids except those with sealing compound. Heat washed jars and lids in clean water before packing with hot food. Some metal lids with sealing compound need boiling; others need only a dip in hot water. Follow manufacturer's directions.

If you use rubber rings, have clean, new rings of the right size for the jars.

Tin Cans

R-enamel—for beets, red berries, red or black cherries, plums, pumpkins, rhubarb, winter squash.

Plain—for all other fruits and vegetables for which canning directions are given in this chapter.

Selecting Fruits and Vegetables for Canning

Choose fresh, firm fruits and young, tender vegetables. Can them quickly, before they lose their freshness. If you must wait, keep them in a cool, airy place. If you buy fruits and vegetables to can, try to get them from a nearby garden or orchard.

It is advisable to use only perfect fruits and vegetables. Sort them for size and ripeness; they cook more evenly that way.

Wash all fruits and vegetables thoroughly, whether or not they are to be pared. Dirt contains some of the bacteria that are hardest to kill. Wash small lots at a time, under running water or through several changes of water. Lift the food out of the water each time so dirt that has been washed off won't go back on the food. Rinse the pan thoroughly between washings. Handle the fruits and vegetables gently to avoid bruising.

Filling Containers

Raw Pack or Hot Pack: Fruits and vegetables may be packed raw into glass jars or tin cans or preheated and packed hot. In this chapter directions for both raw and hot packs are given for most of the foods.

Raw food should be packed into the container tightly because it shrinks during processing. Pack hot food fairly loosely; it should be at or near boiling temperature when it is packed.

There should be enough syrup, water or juice to fill in around the solid food in the container and to cover the food. A quart glass jar or a No. 2½ tin can takes from ½ to 1½ cups of liquid.

Removing air bubbles from the filled containers will help prevent liquid from falling below the level of the solid food during processing; food at the top of the container tends to darken if not covered with liquid.

To remove bubbles, work the blade of a table knife down the sides of the containers. Then add more liquid if needed to cover the food; however, be sure to leave space at the top as directed below.

Raw Pack: Put cold, raw fruits into container and cover with boiling-hot syrup, juice or water. Press tomatoes down in the containers so they are covered with their own juice; add no liquid.

Hot Pack: Heat fruits in syrup, water or steam, or extracted juice before packing. Juicy fruits and tomatoes may be preheated without added liquid and packed in the juice that cooks out.

Head Space: With only a few exceptions, some space should be left between the packed food and the closure. The amount of head space to allow is given in the detailed directions for canning each food.

Exhausting and Sealing Tin Cans

Tin cans are sealed before processing. The temperature of the food in the cans must be 170° F. or higher when the cans are sealed. This drives out air so that there will be a good vacuum in the can after processing and cooling. Removal of air also helps prevent discoloring of canned food and change in flavor.

Food packed raw must be heated in the cans (exhausted) before the cans are sealed. Food packed hot may be sealed without further heating if you are sure the temperature of the food has not dropped below 170° F. To make sure, test with a thermometer, placing the bulb at the center of the can. If the thermometer registers lower than 170°, or if you do not make this test, exhaust the cans.

To exhaust, place open, filled cans on a rack in a kettle in which there is enough boiling water to come to about 2 inches below the tops of the cans. Cover the kettle. Bring the water back to boiling. Boil until a thermometer inserted at the center of the can registers 170° F.—or for the length of time given in the directions for the fruit or vegetable you are canning.

Remove cans from the water one at a time, and add boiling packing liquid or water if necessary to fill to the proper head space. Place a clean lid on the filled can. Seal at once.

Cooling

Glass Jars: As you take jars from the canner, complete the seals at once unless the closure is of a self-sealing type. If liquid has boiled out in the processing, do not open the jar to add more. Seal it just as it is.

Cool jars top side up. Give each jar enough room to let air get at all sides. Never set a hot jar on a cold surface; instead, set the jars on a folded cloth or on a rack. Keep hot jars away from drafts, but don't slow cooling by covering them.

Tin Cans: Put tin cans in cold, clean water to cool them; change the water as needed to cool the cans quickly. Take the cans out of the water while they are still warm so they will dry in the air. If you stack the cans, stagger them so that the air can get around them.

Storing

Canned food should be kept dry and cool, but it should not be subjected to freezing.

Dampness may corrode tin cans and metal lids of glass jars, eventually causing leakage.

Warmth may cause canned food to lose quality. Hot pipes behind a wall sometimes make a shelf or closet too warm for storing food.

Freezing may crack a jar or break a seal and let in bacteria that will cause spoilage.

Sweetening Fruit

Sugar helps canned fruit hold its shape, color and flavor. Directions for canning most fruits call for sweetening in the form of sugar syrup. For very juicy fruit packed hot, use sugar without added liquid.

To Make Sugar Syrup: Mix sugar with water or with juice extracted from some of the fruit.

Type of Syrup	Sugar	Water or Juice	Yield of Syrup
	Cups	Cups	Cups
Thin	2	4	5
Medium	3	4	5½
Heavy	4¾	4	6½

Boil the sugar and water or fruit juice together for 5 minutes. Skim if necessary.

To Extract Juice: Crush thoroughly ripe, sound juicy fruit. Heat to simmering (185° to 210° F.) over low heat. Strain through jelly bag or other cloth.

To Add Sugar Direct to Fruit: For juicy fruit to be packed hot, add about ½ cup of sugar to each quart of raw, prepared fruit. Heat to simmering over low heat. Pack fruit in the juice that cooks out.

To Add Sweetening Other Than Sugar: You can use light corn syrup or mild-flavored honey to replace as much as half the sugar called for in canning fruit. Do not use brown sugar or molasses, sorghum or other strong-flavored syrups; their flavor overpowers the fruit flavor and they may darken the fruit.

You may can fruit without sweetening—in its own juice, in extracted juice or in water. Sugar is not needed to prevent spoilage; processing is the same for unsweetened fruit as for sweetened.

How to Determine Yield of Canned Fruit From Fresh

The number of quarts of canned food you can get from a given quantity of fresh fruit depends on the quality, variety, maturity and size of the fruit, whether it is canned whole or in halves or slices, and whether a raw pack or hot pack is used.

One quart of canned food requires the following amounts of fresh fruit or tomatoes as purchased or picked:

	Pounds
Apples	2½ to 3
Berries, except strawberries	1½ to 3 (1 to 2 quart boxes)
Cherries (canned unpitted)	2 to 2½
Peaches	2 to 3
Pears	2 to 3
Plums	1½ to 2½
Tomatoes	2½ to 3½

In one pound there are about four medium apples, peaches or tomatoes; three medium pears; twelve medium plums.

Processing in Boiling-Water Bath

Directions: Put sealed, filled glass jars or tin cans into a canner containing hot or boiling water. For raw pack in glass jars, have water in the canner hot but not boiling; for all other packs, have water boiling.

Add boiling water if needed to bring the water an inch or two over the tops of the containers; do not pour the boiling water directly on glass jars. Put a cover on the canner.

When the water in the canner comes to a rolling boil, start to count the processing time. Boil gently and steadily for the processing time recommended for the food you are canning. Add boiling water during processing, if needed, to keep containers covered.

Remove the containers from the canner immediately when the processing time is up.

Processing Times: Processing times recommended in the detailed directions are only for foods prepared and packed as specified.

If you live at an altitude less than 1,000 feet above sea level, use the processing times as given in the chart on page 146.

At altitudes of 1,000 feet or more, you must process longer in a boiling-water bath. (See chart.)

HOW TO CAN VARIOUS FRUITS, TOMATOES, PICKLED VEGETABLES

Apples

Pare and core apples; cut in pieces. To keep fruit from darkening, drop the pieces into water containing 2 tablespoons each of salt and vinegar per gallon. Drain, then boil 5 minutes in thin syrup or water.

In Glass Jars: Pack hot fruit to ½ inch of top. Cover with hot syrup or water, leaving a ½-inch space at the top of the jar. Adjust the jar lids. Process in a boiling water bath (212° F.) as follows:

Pint jars	15 minutes
Quart jars	20 minutes

Altitude	Increase in processing time if the time called for is 20 minutes or less	more than 20 minutes
1,000 feet	1 minute	2 minutes
2,000 feet	2 minutes	4 minutes
3,000 feet	3 minutes	6 minutes
4,000 feet	4 minutes	8 minutes
5,000 feet	5 minutes	10 minutes
6,000 feet	6 minutes	12 minutes
7,000 feet	7 minutes	14 minutes
8,000 feet	8 minutes	16 minutes
9,000 feet	9 minutes	18 minutes
10,000 feet	10 minutes	20 minutes

As soon as you remove the jars from the canner, complete the seals if closures are not of the self-sealing type.

In Tin Cans: Pack hot fruit to ¼ inch of top. Fill to top with hot syrup or water. Exhaust to 170° F. (about 10 minutes) and seal the cans. Process in a boiling-water bath (212° F.) as follows:

No. 2 cans	10 minutes
No. 2½ cans	10 minutes

Applesauce

Make applesauce, sweetened or unsweetened. Heat to simmering (185°-210° F.), stirring to keep the applesauce from sticking to pan.

In Glass Jars: Pack hot applesauce to ¼ inch of top. Adjust the lids. Process in a boiling-water bath (212° F.) as follows:

Pint jars	10 minutes
Quart jars	10 minutes

As soon as you remove the jars from the canner, complete the seals if closures are not of the self-sealing type.

In Tin Cans: Pack hot applesauce to top. Exhaust to 170° F. (about 10 minutes) and seal the cans. Process in a boiling-water bath (212° F.) as follows:

No. 2 cans	10 minutes
No. 2½ cans	10 minutes

Berries, Except Strawberries

Raw Pack: Wash berries and drain well.

In Glass Jars: Fill jars to ½ inch of top. For a full pack, shake the berries down while filling the jars. Cover with boiling syrup, leaving ½-inch space at the top. Adjust the lids. Process in a boiling-water bath (212° F.) as follows:

Pint jars	10 minutes
Quart jars	15 minutes

As soon as you remove the jars from the canner, complete the seals if closures are not of the self-sealing type.

In Tin Cans: Fill cans to ¼ inch of top. For a full pack, shake the berries down while filling the cans. Fill to the top with boiling syrup. Exhaust to 170° F. (about 10 minutes) and seal the cans. Process in a boiling-water bath (212°) as follows:

No. 2 cans	15 minutes
No. 2½ cans	20 minutes

Hot Pack (for firm berries): Wash berries and drain well. Add ½ cup sugar to each quart fruit. Cover the pan and bring to a boil; shake the pan to keep the berries from sticking.

In Glass Jars: Pack hot berries to ½ inch of top. Adjust the jar lids. Process in a boiling-water bath (212° F.) as follows:

Pint jars	10 minutes
Quart jars	15 minutes

As soon as you remove the jars from the canner, complete the seals if closures are not of the self-sealing type.

In Tin Cans: Pack hot berries to top. Exhaust to 170° F. (about 10 minutes) and seal the cans. Process in a boiling-water bath (212° F.) as follows:

No. 2 cans	15 minutes
No. 2½ cans	20 minutes

Cherries

Raw Pack: Wash cherries; remove pits, if desired.

In Glass Jars: Fill jars to ½ inch of top. For a full pack, shake the cherries down while filling the jars. Cover with a boiling syrup, leaving ½-inch space at the top. Adjust the lids. Process in a boiling-water bath (212° F.) as follows:

Pint jars	20 minutes
Quart jars	25 minutes

As soon as you remove the jars from the canner, complete the seals if closures are not of the self-sealing type.

In Tin Cans: Fill cans to ¼ inch of top. For a full pack, shake the cherries down while filling the cans. Fill to the top with boiling syrup. Exhaust to 170° F. (about 10 minutes) and seal the cans. Process in a boiling-water bath (212° F.) as follows:

No. 2 cans	20 minutes
No. 2½ cans	25 minutes

Hot Pack: Wash cherries; remove pits, if desired. Add ½ cup sugar to each quart fruit. Add a little water to unpitted cherries to keep them from sticking while heating. Cover the pan and bring to a boil.

In Glass Jars: Pack hot to ½ inch of top. Adjust the jar lids. Process in a boiling-water bath (212° F.) as follows:

Pint jars	10 minutes
Quart jars	15 minutes

As soon as you remove the jars from the canner, complete the seals if closures are not of the self-sealing type.

In Tin Cans: Pack hot to top of cans. Exhaust to 170° F. (about 10 minutes) and seal the cans. Process in a boiling-water bath (212° F.) as follows:

No. 2 cans	15 minutes
No 2½ cans	20 minutes

Fruit Juices

Wash; remove pits, if desired, and crush fruit. Heat to simmering (185°–210° F.). Strain through a cloth bag. Add sugar, if desired—about one cup to one gallon of juice. Reheat to simmering.

In Glass Jars: Fill jars to top with hot juice. Adjust the jar lids. Process in a boiling-water bath (212° F.) as follows:

Pint jars	5 minutes
Quart jars	5 minutes

As soon as you remove the jars from the canner, complete the seals if closures are not of the self-sealing type.

In Tin Cans: Fill cans to top with hot juice. Seal at once. Process in a boiling-water bath (212° F.) as follows:

No. 2 cans	5 minutes
No 2½ cans	5 minutes

Peaches

Wash peaches and remove skins. Dip the fruit in boiling water, then quickly in cold water, to make peeling easier. Cut the peaches in halves; remove the pits. Slice, if desired. To prevent the fruit from darkening during preparation, drop it into water containing 2 tablespoons each of salt and vinegar per gallon. Drain just before heating or packing raw.

Raw Pack: Prepare peaches as directed above.

In Glass Jars: Pack raw fruit to ½ inch of top. Cover with boiling syrup, leaving ½-inch space at the top of the jar. Adjust the jar lids. Process in a boiling-water bath (212° F.) as follows:

Pint jars	25 minutes
Quart jars	30 minutes

As soon as you remove the jars from the canner, complete the seals if closures are not of the self-sealing type.

In Tin Cans: Pack raw fruit to ¼ inch of top. Fill to the top with boiling syrup. Exhaust to 170° F. (about 10 minutes) and seal the cans. Process in a boiling-water bath (212° F.) as follows:

No. 2 cans	30 minutes
No. 2½ cans	35 minutes

Hot Pack: Prepare peaches as directed above. Heat the peaches through in hot syrup. If the fruit is very juicy you may heat it with sugar, adding no liquid.

In Glass Jars: Pack hot fruit to ½ inch of top. Cover with boiling liquid, leaving ½-inch space at the top of the jar. Adjust the jar lids. Process in a boiling-water bath (212° F.) as follows:

Pint jars	20 minutes
Quart jars	25 minutes

As soon as you remove the jars from the canner, complete the seals if the closures are not of the self-sealing type.

In Tin Cans: Pack hot fruit to ¼ inch of top. Fill to the top with boiling liquid. Exhaust to 170° F. (about 10 minutes) and seal the cans. Process in a boiling-water bath (212° F.) as follows:

No. 2 cans	25 minutes
No. 2½ cans	30 minutes

Pears

Wash pears. Peel, cut in halves and core. Continue as with peaches, either raw pack or hot pack.

Plums

Wash plums. To can whole, prick the skins. Freestone varieties may be halved and pitted.

Raw Pack: Prepare plums as directed above.

In Glass Jars: Pack raw fruit to ½ inch of top. Cover with boiling syrup, leaving ½-inch space at the top of the jar. Adjust the jar lids. Process in a boiling-water bath (212° F.) as follows:

Pint jars	20 minutes
Quart jars	25 minutes

As soon as you remove the jars from the canner, complete the seals if the closures are not of the self-sealing type.

In Tin Cans: Pack raw fruit to ¼ inch of top. Fill to the top with boiling syrup. Exhaust to 170° F. (about 10 minutes) and seal the cans. Process in a boiling-water bath (212° F.) as follows:

No. 2 cans	15 minutes
No. 2½ cans	20 minutes

Hot Pack: Prepare plums as directed on this page. Heat to boiling in syrup or juice. If the fruit is very juicy you may heat it with sugar, adding no liquid.

In Glass Jars: Pack hot fruit to ½ inch of top. Cover with boiling liquid, leaving ½-inch space at the top of the jar. Adjust the jar lids. Process in a boiling-water bath (212° F.) as follows:

Pint jars	20 minutes
Quart jars	25 minutes

As soon as you remove the jars from the canner, complete the seals if the closures are not of the self-sealing type.

In Tin Cans: Pack hot fruit to ¼ inch of top. Fill to the top with boiling liquid. Exhaust to 170° F.) as follows:

No. 2 cans	15 minutes
No. 2½ cans	20 minutes

Rhubarb

Wash rhubarb and cut into ½-inch pieces. Add

½ cup sugar to each quart rhubarb and let it stand to draw out the juice. Bring to boiling.
In Glass Jars: Pack hot to ½ inch of top. Adjust the lids. Process in a boiling-water bath (212° F.) as follows:

Pint jars	10 minutes
Quart jars	10 minutes

As soon as you remove the jars from the canner, complete the seals if closures are not of the self-sealing type.
In Tin Cans: Pack hot to top of cans. Exhaust to 170° F. (about 10 minutes) and seal the cans. Process in a boiling-water bath (212° F.) as follows:

No. 2 cans	10 minutes
No. 2½ cans	10 minutes

Sauerkraut

Heat well-fermented sauerkraut to simmering (185°–210° F.). Do not boil.
In Glass Jars: Pack hot kraut to ½ inch of top. Cover with hot juice, leaving ½-inch space at the top of the jar. Adjust the jar lids. Process in a boiling-water bath (212° F.) as follows:

Pint jars	15 minutes
Quart jars	20 minutes

As soon as you remove the jars from the canner, complete the seals if closures are not of the self-sealing type.
In Tin Cans: Pack hot kraut to ¼ inch of top. Fill to top with hot juice. Exhaust to 170° F. (about 10 minutes) and seal the cans. Process in a boiling-water bath (212° F.) as follows:

No. 2 cans	20 minutes
No 2½ cans	25 minutes

Tomatoes

Use only perfect, ripe tomatoes. To loosen the skins, dip into boiling water for about ½ minute; then dip quickly into cold water. Cut out the stem ends and peel the tomatoes.

Raw Pack: Leave tomatoes whole or cut in halves or quarters.
In Glass Jars: Pack tomatoes to ½ inch of top, pressing gently to fill spaces. Add no water. Add ½ teaspoon salt to pints; one teaspoon to quarts. Adjust the lids. Process in a boiling-water bath (212° F.) as follows:

Pint jars	35 minutes
Quart jars	45 minutes

As soon as you remove the jars from the canner, complete the seals if closures are not of the self-sealing type.
In Tin Cans: Pack tomatoes to top of cans, pressing gently to fill spaces. Add no water. Add ½ teaspoon salt to No. 2 cans; one teaspoon to No. 2½. Exhaust to 170° F. (about 15 minutes) and seal the cans. Process in a boiling-water bath (212° F.) as follows:

No. 2 cans	45 minutes
No 2½ cans	55 minutes

Hot Pack: Quarter peeled tomatoes. Bring to a boil; stir to keep the tomatoes from sticking.
In Glass Jars: Pack boiling-hot tomatoes to ½ inch of top. Add ½ teaspoon salt to pints; one teaspoon to quarts. Adjust the jar lids. Process in a boiling-water bath (212° F.) as follows:

Pint jars	10 minutes
Quart jars	10 minutes

As soon as you remove the jars from the canner, complete the seals if closures are not of the self-sealing type.
In Tin Cans: Pack boiling-hot tomatoes to ¼ inch of top. Add no water. Add ½ teaspoon salt to No. 2 cans; one teaspoon to No. 2½ cans. Exhaust to 170° F. (about 10 minutes) and seal the cans. Process in a boiling-water bath (212° F.) as follows:

No. 2 cans	10 minutes
No. 2½ cans	10 minutes

Tomato Juice

Use ripe, juicy tomatoes. Wash, remove the stem ends, cut into pieces. Simmer until softened, stirring often. Put through a strainer. Add one teaspoon salt to each quart juice. Reheat at once just to boiling.

In Glass Jars: Fill jars with boiling-hot juice to ¼ inch of top. Adjust the jar lids. Process in a boiling-water bath (212° F.) as follows:

Pint jars	10 minutes
Quart jars	10 minutes

As soon as you remove the jars from the canner, complete the seals if closures are not of the self-sealing type.

In Tin Cans: Fill cans to top with boiling-hot juice. Seal the cans at once. Process in a boiling-water bath (212° F.) as follows:

No. 2 cans	15 minutes
No 2½ cans	15 minutes

Asparagus

Raw Pack: Wash asparagus; trim off scales and tough ends and wash again. Cut into one-inch pieces.

In Glass Jars: Pack asparagus as tightly as possible without crushing to ½ inch of top. Add ½ teaspoon salt to pints; one teaspoon to quarts. Cover with boiling water, leaving ½-inch space at the top of jar. Adjust the jar lids. Process in a pressure canner at 10 pounds pressure (240° F.) as follows:

Pint jars	25 minutes
Quart jars	30 minutes

As soon as you remove the jars from the canner, complete the seals if closures are not of the self-sealing type.

In Tin Cans: Pack asparagus as tightly as possible without crushing to ¼ inch of top. Add ½ teaspoon salt to No. 2 cans; one teaspoon to No. 2½ cans. Fill to the top with boiling water. Exhaust to 170° F. (about 10 minutes) and seal the cans. Process in a pressure canner at 10 pounds pressure (240° F.) as follows:

No. 2 cans	20 minutes
No 2½ cans	20 minutes

Hot Pack: Wash asparagus; trim off scales and tough ends and wash again. Cut into one-inch pieces. Cover with boiling water. Boil 2 or 3 minutes.

In Glass Jars: Pack hot asparagus loosely to ½ inch of top. Add ½ teaspoon salt to pints; one teaspoon to quarts. Cover with boiling-hot cooking liquid, or if liquid contains grit use boiling water. Leave ½-inch space at the top of the jar. Adjust the jar lids. Process in a pressure canner at 10 pounds pressure (240° F.) as follows:

Pint jars	25 minutes
Quart jars	30 minutes

As soon as you remove the jars from the canner, complete the seals if closures are not of the self-sealing type.

In Tin Cans: Pack hot asparagus loosely to ¼ inch of top. Add ½ teaspoon salt to No. 2 cans; one teaspoon to No. 2½ cans. Fill to the top with boiling-hot cooking liquid, or if liquid contains grit use boiling water. Exhaust to 170° F. (about 10 minutes) and seal the cans. Process in a pressure canner at 10 pounds pressure (240° F.) as follows:

No. 2 cans	20 minutes
No. 2½ cans	20 minutes

Beans, Snap

Raw Pack: Wash beans. Trim ends; cut into one-inch pieces.

In Glass Jars: Pack raw beans tightly to ½ inch of top. Add ½ teaspoon salt to pints; one teaspoon to quarts. Cover with boiling water, leaving ½-inch space at the top of the jar. Adjust the jar lids. Process in a pressure canner at 10 pounds pressure (240° F.) as follows:

Pint jars .... 20 minutes
Quart jars .... 25 minutes

As soon as you remove the jars from the canner, complete the seals if closures are not of the self-sealing type.

In Tin Cans: Pack raw beans tightly to ¼ inch of top. Add ½ teaspoon salt to No. 2 cans; one teaspoon to No. 2½ cans. Fill to the top with boiling water. Exhaust to 170° F. (about 10 minutes) and seal the cans. Process in a pressure canner at 10 pounds pressure (240° F.) as follows:

No. 2 cans .... 25 minutes
No. 2½ cans .... 30 minutes

Hot Pack: Wash beans. Trim the ends; cut into one-inch pieces. Cover with boiling water; boil 5 minutes.

In Glass Jars: Pack hot beans loosely to ½ inch of top. Add ½ teaspoon salt to pints; one teaspoon to quarts. Cover with boiling-hot cooking liquid, leaving ½-inch space at the top of the jar. Adjust the jar lids. Process in a pressure canner at 10 pounds pressure (240° F.) as follows:

Pint jars .... 20 minutes
Quart Jars .... 25 minutes

As soon as you remove the jars from the canner, complete the seals if closures are not of the self-sealing type.

In Tin Cans: Pack hot beans loosely to ¼ inch of top. Add ½ teaspoon salt to No. 2 cans; one teaspoon to No. 2½ cans. Fill to the top with boiling-hot cooking liquid. Exhaust to 170° F. (about 10 minutes) and seal the cans. Process in a pressure canner at 10 pounds pressure (240° F.) as follows:

No. 2 cans .... 25 minutes
No 2½ cans .... 30 minutes

Beets

Sort beets for size. Cut off the tops, leaving an inch of stem. Also leave the root. Wash the beets. Cover with boiling water and boil until the skins slip easily–15 to 25 minutes, depending on size. Skin and trim. Leave baby beets whole. Cut medium or large beets in ½-inch cubes or slices; halve or quarter very large slices.

In Glass Jars: Pack hot beets to ½ inch of top. Add ½ teaspoon salt to pints; one teaspoon to quarts. Cover with boiling water, leaving ½-inch space at the top of the jar. Adjust the jar lids. Process in a pressure canner at 10 pounds pressure (240° F.) as follows:

Pint jars .... 30 minutes
Quart jars .... 35 minutes

As soon as you remove the jars from the canner, complete the seals if closures are not of the self-sealing type.

In Tin Cans: Pack hot beets to ¼ inch of top. Add ½ teaspoon salt to No. 2 cans; one teaspoon to No. 2½ cans. Fill to the top with boiling water. Exhaust to 170° F. (about 10 minutes) and seal the cans. Process in a pressure canner at 10 pounds pressure (240° F.) as follows:

No. 2 cans .... 30 minutes
No. 2½ cans .... 30 minutes

Beets, Pickled

Cut off beet tops, leaving one inch of stem. Also leave the root. Wash the beets, cover with boiling water and cook until tender. Remove skins and slice the beets. For pickling syrup, use 2 cups vinegar (or 1½ cups vinegar and ½ cup water) to 2 cups sugar. Heat to boiling.

Pack beets in glass jars to ½ inch of top. Add ½ teaspoon salt to pints, one teaspoon to quarts. Cover with boiling syrup, leaving ½-inch space at the top of the jar. Adjust the jar lids. Process in a boiling-water bath (212° F.) as follows:

Pint jars .... 30 minutes
Quart jars .... 30 minutes

As soon as you remove the jars from the canner,

complete the seals if closures are not of the self-sealing type.

Carrots
Raw Pack: Wash and scrape carrots. Slice or dice.
In Glass Jars: Pack raw carrots tightly into clean jars, to one inch of top of jar. Add ½ teaspoon salt to pints; one teaspoon to quarts. Fill the jar to the top with boiling water. Adjust the jar lids. Process in a pressure canner at 10 pounds pressure (240° F.) as follows:

Pint jars . . . . 25 minutes
Quart jars . . . . 30 minutes

As soon as you remove the jars from the canner, complete the seals if closures are not of the self-sealing type.
In Tin Cans: Pack raw carrots tightly into cans to ½ inch of top. Add ½ teaspoon salt to No. 2 cans; one teaspoon to No. 2½ cans. Fill the cans to the top with boiling water. Exhaust to 170° F. (about 10 minutes) and seal the cans. Process in a pressure canner at 10 pounds pressure (240° F.) as follows:

No. 2 cans . . . . 25 minutes
No. 2½ cans . . . . 30 minutes

Hot Pack: Wash and scrape carrots. Slice or dice. Cover with boiling water and bring to boil.
In Glass Jars: Pack hot carrots to ½ inch of top. Add ½ teaspoon salt to pints; one teaspoon to quarts. Cover with boiling-hot cooking liquid, leaving ½-inch space at the top of the jar. Adjust the jar lids. Process in a pressure canner at 10 pounds pressure (240° F.) as follows:

Pint jars . . . . 25 minutes
Quart jars . . . . 30 minutes

As soon as you remove the jars from the canner, complete the seals if closures are not of the self-sealing type.
In Tin Cans: Pack hot carrots to ¼ inch of top. Add ½ teaspoon salt to No. 2 cans; one teaspoon to No. 2½ cans. Fill with boiling-hot cooking liquid. Exhaust to 170° F. (about 10 minutes) and seal the cans. Process in a pressure canner at 10 pounds pressure (240° F.) as follows:

No. 2 cans . . . . 20 minutes
No. 2½ cans . . . . 25 minutes

Corn, Whole-Kernel
Raw Pack: Husk corn and remove silk. Wash. Cut from the cob at about two-thirds the depth of the kernel.
In Glass Jars: Pack corn to one inch of top; do not shake or press down. Add ½ teaspoon salt to pints; one teaspoon to quarts. Fill to the top with boiling water. Adjust the jar lids. Process in a pressure canner at 10 pounds pressure (240° F.) as follows:

In pint jars . . . . 55 minutes
Quart jars . . . . 85 minutes

As soon as you remove the jars from the canner, complete the seals if closures are not of the self-sealing type.
In Tin Cans: Pack corn to ½ inch of top; do not shake or press down. Add ½ teaspoon salt to No. 2 cans; one teaspoon to No. 2½ cans. Fill to the top with boiling water. Exhaust to 170° F. (about 10 minutes) and seal the cans. Process in a pressure canner at 10 pounds pressure (240° F.) as follows:

No. 2 cans . . . . 60 minutes
No. 2½ cans . . . . 60 minutes

Hot Pack: Husk corn and remove silk. Wash. Cut from the cob at about two-thirds the depth of the kernel. To each quart of corn add one pint boiling water. Heat to boiling.
In Glass Jars: Pack hot corn to one inch of top and cover with boiling-hot cooking liquid, leaving one-inch space at top of jar. Or fill to one inch of top with a mixture of corn and liquid. Add ½ teaspoon salt to pints; one teaspoon to quarts. Adjust the jar lids. Process in a pressure

canner at 10 pounds pressure (240° F.) as follows:

Pint jars .... 55 minutes
Quart jars .... 85 minutes

As soon as you remove the jars from the canner, complete the seals if closures are not of the self-sealing type.

In Tin Cans: Pack hot corn to ½ inch of top and fill to top with boiling-hot cooking liquid. Or fill to the top with a mixture of corn and liquid. Add ½ teaspoon salt to No. 2 cans; one teaspoon to No. 2½ cans. Exhaust to 170° F. (about 10 minutes) and seal the cans. Process in a pressure canner at 10 pounds pressure (240° F.) as follows:

No. 2 cans .... 60 minutes
No. 2½ cans .... 60 minutes

Peas, Fresh Green

Raw Pack: Shell and wash peas.

In Glass Jars: Pack peas to one inch of top; do not shake or press down. Add ½ teaspoon salt to pints; one teaspoon to quarts. Cover with boiling water, leaving one-inch space at the top of the jar. Adjust the jar lids. Process in a pressure canner at 10 pounds pressure (240° F.) as follows:

Pint jars .... 40 minutes
Quart jars .... 40 minutes

As soon as you remove the jars from the canner, complete the seals if closures are not of the self-sealing type.

In Tin Cans: Pack peas to ¼ inch of top; do not shake or press down. Add ½ teaspoon salt to No. 2 cans; one teaspoon to No. 2½ cans. Fill to the top with boiling water. Exhaust to 170° F. (about 10 minutes) and seal the cans. Process at 10 pounds pressure (240° F.) as follows:

No. 2 cans .... 30 minutes
No. 2½ cans .... 35 minutes

Hot Pack: Shell and wash peas. Cover with boiling water. Bring to boil.

In Glass Jars: Pack hot peas loosely to one inch of top. Add ½ teaspoon salt to pints; one teaspoon to quarts. Cover with boiling water, leaving one-inch space at the top of the jar. Adjust the jar lids. Process in a pressure canner at 10 pounds pressure (240° F.) as follows:

Pint jars .... 40 minutes
Quart jars .... 40 minutes

As soon as you remove the jars from the canner, complete the seals if closures are not of the self-sealing type.

In Tin Cans: Pack hot peas loosely to ¼ inch of top. Add ½ teaspoon salt to No. 2 cans; one teaspoon to No. 2½ cans. Fill to the top with boiling water. Exhaust to 170° F. (about 10 minutes) and seal the cans. Process at 10 pounds pressure (240° F.) as follows:

No. 2 cans .... 30 minutes
No. 2½ cans .... 35 minutes

Spinach (and Other Greens)

Can only freshly picked, tender spinach. Pick over and wash thoroughly. Cut out tough stems and midribs. Place about 2½ pounds of spinach in a cheesecloth bag or covered pot and steam about 10 minutes or until well wilted.

In Glass Jars: Pack hot spinach loosely to ½ inch of top. Add ¼ teaspoon salt to pints; ½ teaspoon to quarts. Cover with boiling water, leaving ½-inch space at the top of the jar. Adjust the jar lids. Process in a pressure canner at 10 pounds pressure (240° F.) as follows:

Pint jars .... 70 minutes
Quart jars .... 90 minutes

As soon as you remove the jars from the canner, complete the seals if closures are not of the self-sealing type.

In Tin Cans: Pack hot spinach loosely to ¼ inch of top. Add ¼ teaspoon salt to No. 2 cans; ½ teaspoon to No. 2½ cans. Fill to the top with boiling water. Exhaust to 170° F. (about 10 minutes) and seal the cans. Process in a pressure canner at 10 pounds pressure (240° F.) as follows:

No. 2 cans	65 minutes
No. 2½ cans	75 minutes

WILD BERRY JAM AND JELLIES

Uncooked Method

Uncooked jam is the easiest of all to make. You can make it from fresh fruit or from fruit that has been frozen. Commercial pectin is used in all these jams. Lemon juice is added to supply extra acids; a higher percentage of sugar is needed to form a jell. Store uncooked jam in the refrigerator or freezer and use within three months.

BERRY JAM UNCOOKED

2 cups crushed berries
1 package powdered pectin
1 cup water
4 cups sugar
juice of 1 lemon

1. Mix berries and sugar. Let stand 20 minutes, stirring occasionally.
2. Combine pectin with the water, bring to a boil and boil one minute, stirring constantly.
3. Add pectin to the berries, sugar and lemon juice. Stir about 2 minutes more.
4. Pour into clean sterilized jars and cover with a lid of aluminum foil.
5. Let stand at room temperature for 24 hours until it congeals. Then refrigerate until it is set.
6. Store in a refrigerator or freezer until used.

Jellies: Jellies, like jams, can also be made uncooked.

Blueberry and the trailing raspberry make excellent jelly. Their seeds are slightly larger and sharper; therefore you may wish to extract the juice and leave the seeds behind.

Extract the juice by heating and crushing a few berries with some water (a cup or less) to start. Then add the remainder of your berries and simmer to extract the most juice possible.

To get a clear juice without sediment, add one cup of cellulose pulp for every 3 cups of crushed berries and stir while cooking. (To make cellulose pulp, place ten white facial tissues in 2 quarts of boiling water, stirring constantly to prevent sticking. Let stand for a minute, then whip with a fork to make a pulp. Pour into a strainer and shake, but do not press, to remove excess water. Makes one cup of cellulose pulp.) Pour hot fruit and cellulose into a jelly bag or several thicknesses of cheesecloth in a strainer or colander. Let the juice drip into a bowl. When cool enough to handle, twist the bag or press the pulp against the side of the colander to extract the last bit of juice. Then you can discard the pulp.

Here is a basic recipe for all uncooked berry jellies:

UNCOOKED BERRY JELLY

3 cups berry juice
4½ cups sugar
1 box powdered pectin
½ cup water

1. Add the sugar to 1½ cups of the berry juice and stir thoroughly.
2. Add powdered pectin slowly to the ½ cup of water and heat almost to boiling, stirring constantly.
3. Pour the pectin mixture into remaining 1½ cups of berry juice and stir until pectin is completely dissolved.
4. Let the pectin mixture stand 15 minutes and stir it occasionally.
5. Combine the juice and pectin mixtures and stir until all the sugar is dissolved.
6. Pour into containers and let stand at room temperature until set, which may be from 6 hours to overnight.
7. Store in a refrigerator or freezer. Use within 3 months.

Here is a recipe for uncooked jelly without pectin:

UNCOOKED CURRANT JELLY

4 cups currant juice
5½ cups sugar

1. Bring juice to a boil.
2. Add sugar to the boiling juice.
3. Heat again to dissolve the sugar.
4. Seal in sterilized jars.

It is not essential to can all vegetables and fruit to keep them until they are eaten. Some can be stored, above the freezing point, outdoors in pits, trenches and root cellars or indoors in a spare room or in your basement.

Many fruits and vegetables can also be kept in a home freezer. If you own a home freezer or intend to purchase one, see freezing instructions that accompany the freezer.

If your plan is to store the vegetables and fruits that lend themselves neither to canning or freezing, then the following chart will give you storage hints, conditions such as temperature and humidity, and the maximum length of storage for each type or commodity.

Meat can also be kept frozen in your home freezer rather than smoked or otherwise preserved. See manufacturer's instructions for methods to suit each cut and animal.

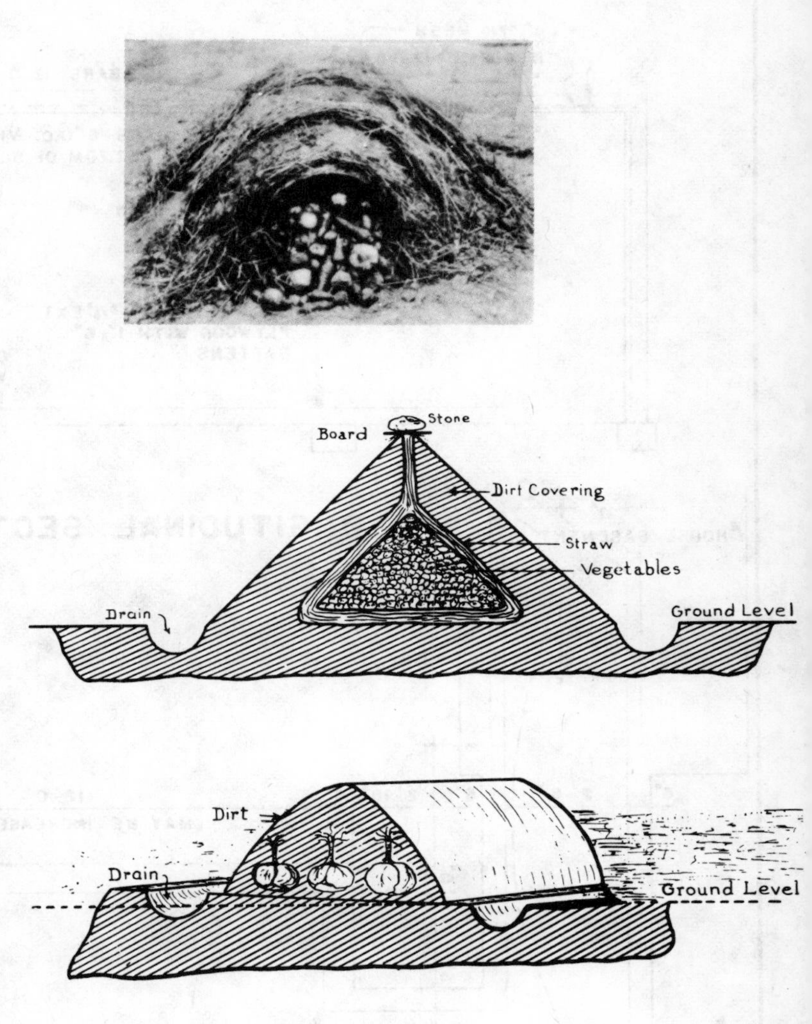

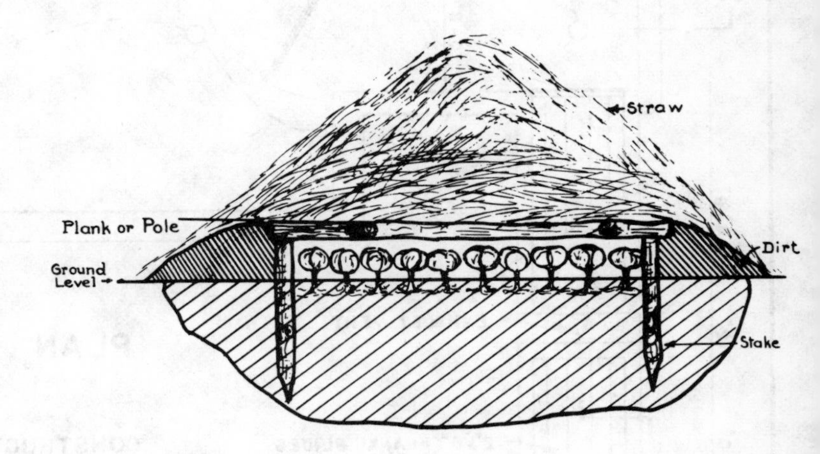

Storing vegetables outdoors: (a) A straw and soil-covered barrel is adequate for keeping a small quantity of vegetables or fruits; (b) cone-shaped pit shows details of construction; (c) cabbages are placed head-down in a long pit; (d) cabbages are placed upright in a trench that is framed with stakes and covered with straw.

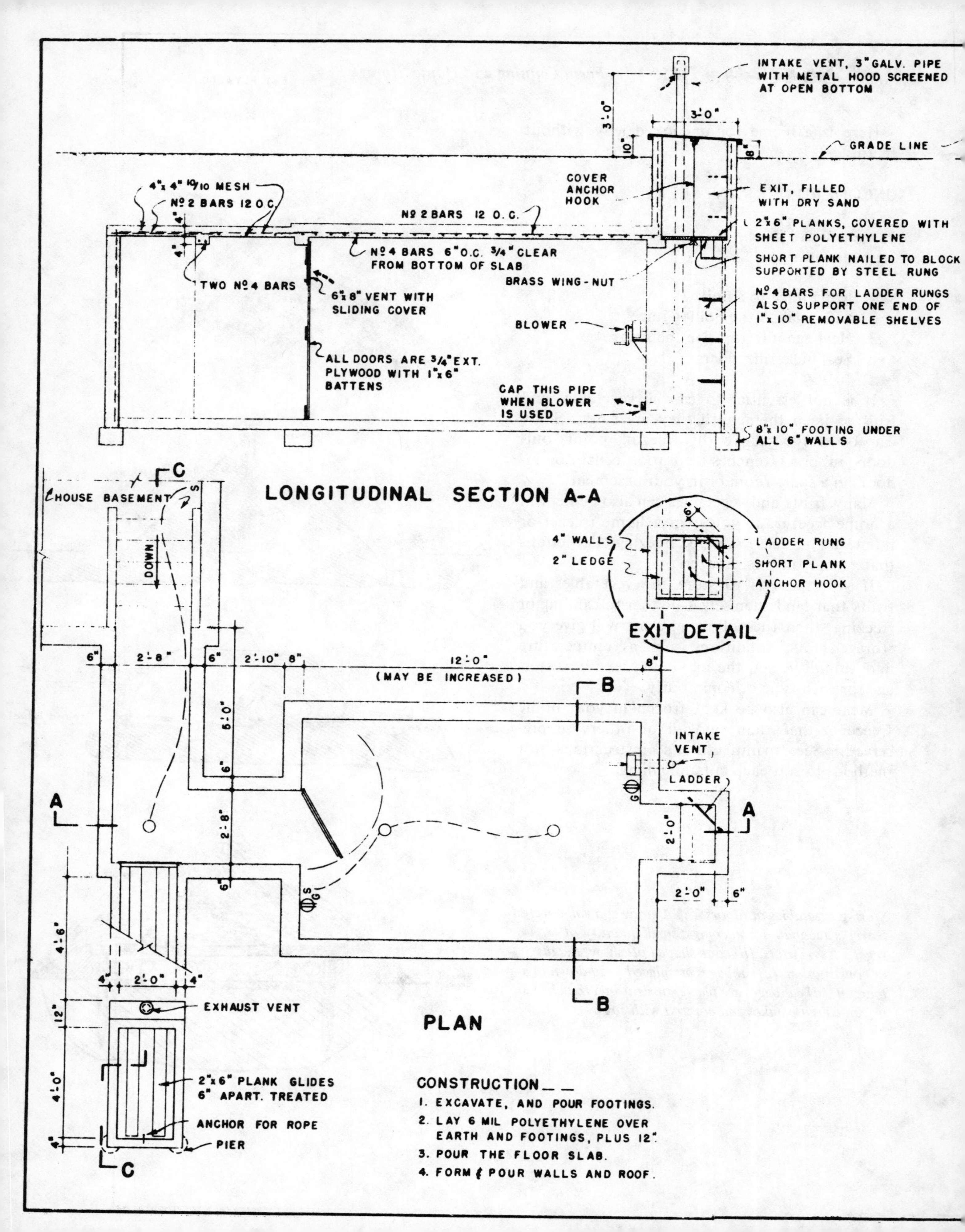
INTAKE VENT, 3" GALV. PIPE WITH METAL HOOD SCREENED AT OPEN BOTTOM
3'-0"
3'-0"
10"
8"
GRADE LINE
4"x 4" 10/10 MESH
No 2 BARS 12 O.C.
No 2 BARS 12 O.C.
COVER ANCHOR HOOK
EXIT, FILLED WITH DRY SAND
2"x 6" PLANKS, COVERED WITH SHEET POLYETHYLENE
No 4 BARS 6" O.C. 3/4" CLEAR FROM BOTTOM OF SLAB
SHORT PLANK NAILED TO BLOCK SUPPORTED BY STEEL RUNG
BRASS WING-NUT
TWO No 4 BARS
No 4 BARS FOR LADDER RUNGS ALSO SUPPORT ONE END OF 1"x 10" REMOVABLE SHELVES
6"x 8" VENT WITH SLIDING COVER
BLOWER
ALL DOORS ARE 3/4" EXT. PLYWOOD WITH 1"x 6" BATTENS
CAP THIS PIPE WHEN BLOWER IS USED
8"x 10" FOOTING UNDER ALL 6" WALLS
LONGITUDINAL SECTION A-A
HOUSE BASEMENT
DOWN
4" WALLS
2" LEDGE
LADDER RUNG
SHORT PLANK
ANCHOR HOOK
EXIT DETAIL
6"
2'-8"
6"
2'-10"
8"
12'-0"
(MAY BE INCREASED)
8"
5'-0"
INTAKE VENT
LADDER
2'-8"
2'-0"
2'-0"
6"
4'-6"
4"
2'-0"
4"
12"
EXHAUST VENT
PLAN
4'-0"
2"x 6" PLANK GLIDES 6" APART. TREATED
ANCHOR FOR ROPE
PIER
CONSTRUCTION
1. EXCAVATE, AND POUR FOOTINGS.
2. LAY 6 MIL POLYETHYLENE OVER EARTH AND FOOTINGS, PLUS 12".
3. POUR THE FLOOR SLAB.
4. FORM & POUR WALLS AND ROOF.

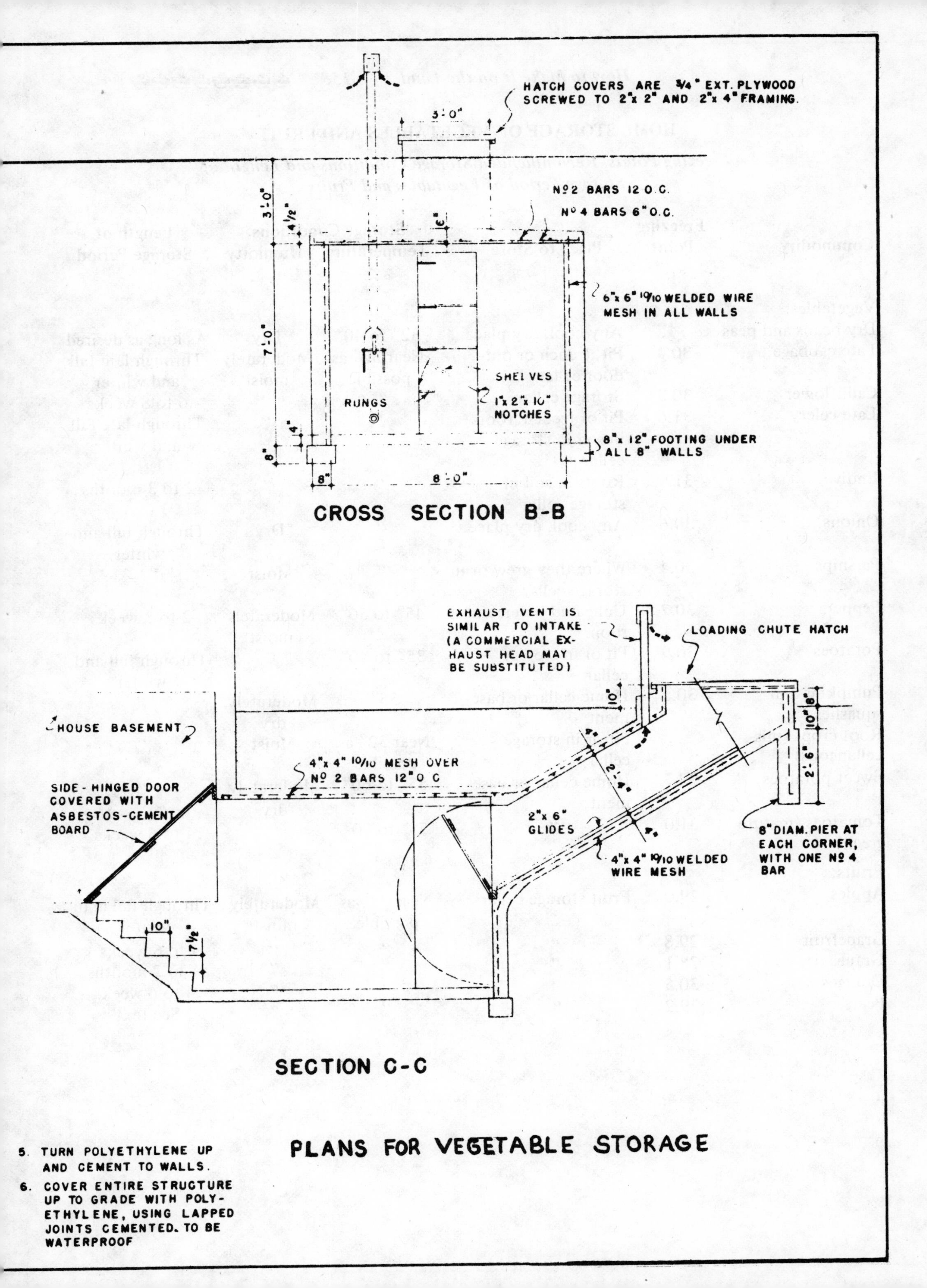
HATCH COVERS ARE 3/4" EXT. PLYWOOD SCREWED TO 2"x 2" AND 2"x 4" FRAMING.
3'-0"
3'-0"
4 1/2"
Nº 2 BARS 12 O.C.
Nº 4 BARS 6" O.C.
6"
6"x 6" 10/10 WELDED WIRE MESH IN ALL WALLS
7'-0"
SHELVES
1"x 2"x 10" NOTCHES
RUNGS
4"
8"
8"
8'-0"
8"x 12" FOOTING UNDER ALL 8" WALLS
CROSS SECTION B-B
EXHAUST VENT IS SIMILAR TO INTAKE (A COMMERCIAL EXHAUST HEAD MAY BE SUBSTITUTED)
LOADING CHUTE HATCH
10"
8"
10"
2'-6"
HOUSE BASEMENT
4"x 4" 10/10 MESH OVER Nº 2 BARS 12" O.C
SIDE-HINGED DOOR COVERED WITH ASBESTOS-CEMENT BOARD
2"x 6" GLIDES
4"x 4" 10/10 WELDED WIRE MESH
8" DIAM. PIER AT EACH CORNER, WITH ONE Nº 4 BAR
10"
7 1/2"
SECTION C-C
PLANS FOR VEGETABLE STORAGE
5. TURN POLYETHYLENE UP AND CEMENT TO WALLS.
6. COVER ENTIRE STRUCTURE UP TO GRADE WITH POLYETHYLENE, USING LAPPED JOINTS CEMENTED TO BE WATERPROOF

HOME STORAGE OF VEGETABLES AND FRUIT

Freezing Points, Recommended Storage Conditions and Length of Storage Period of Vegetables and Fruits

Commodity	Freezing Point	Place to Store	Storage Conditions: Temperature	Humidity	Length of Storage Period
	°F.		°F.		
Vegetables:					
Dry beans and peas	. . .	Any cool, dry place	32° to 40°	Dry	As long as desired
Late cabbage	30.4	Pit, trench or outdoor cellar	Near 30° as possible	Moderately moist	Through late fall and winter
Cauliflower	30.3	Storage cellar	"	"	6 to 8 weeks
Late celery	31.6	Pit or trench; roots in soil in storage cellar	"	"	Through late fall and winter
Endive	31.9	Roots in soil in storage cellar	"	"	2 to 3 months
Onions	30.6	Any cool, dry place	"	Dry	Through fall and winter
Parsnips	30.4	Where they grew or in storage cellar	"	Moist	"
Peppers	30.7	Unheated basement or room	45° to 50°	Moderately moist	2 to 3 weeks
Potatoes	30.9	Pit or in storage cellar	35° to 40°	"	Through fall and winter
Pumpkins and squashes	30.5	Home cellar or basement	55°	Moderately dry	"
Root crops (miscellaneous)	. . .	Pit or in storage cellar	Near 32° as possible	Moist	"
Sweet potatoes	29.7	Home cellar or basement	55° to 60°	Moderately dry	"
Tomatoes (mature green)	31.0	"	55° to 70°	"	4 to 6 weeks
Fruits:					
Apples	29.0	Fruit storage cellar	Near 32° as possible	Moderately moist	Through fall and winter
Grapefruit	29.8	"	"	"	4 to 6 weeks
Grapes	28.1	"	"	"	1 to 2 months
Oranges	30.5	"	"	"	4 to 6 weeks
Pears	29.2	"	"	"	See text

16

Preserving Meats for Year-Round Use

A YEAR-ROUND SUPPLY OF MEAT

In 1948 the production of home freezers in this country reached a stage that brought the cost of a unit within reach of the average householder. Until then farmwives and other people who owned enough land to produce most of the food their families needed for the year preserved their annual supplies of meat, poultry, fish, vegetables and fruit pretty much the way used by the earliest settlers of this land. In the case of meat, to which this particular chapter is devoted, that meant salting, smoking, drying, putting it down into deep stone crocks and covering it with layers of its own fat or, during the winter months, placing it in linen bags and hanging it outdoors, high enough off the ground to protect it from wild scavengers.

Since 1948, when the home freezer industry produced 690,000 units, millions of economy-minded families have purchased freezers for storing their annual supply of home-grown food. These freezers can even be found in remote areas which are still without electricity, operated by home power plants that can be purchased for as little as $600.

In spite of this fact, many housewives who can easily afford the conveniences of a freezer still preserve their home-grown products in the old-fashioned way because they feel it is more economical, and the food tastes better. Also, a lot of people enjoy the feeling of self-reliance achieved by doing without freezers and other modern gimmicks.

There are eight basic methods for preserving meat: dry salting, corning, smoking, crock preserving, pickling, drying, natural freezing and home freezing. Meats covered here include pork, veal, beef, lamb and poultry.

Pork

COUNTRY CURED HAM

150 pounds of fresh hams (approximate)
3½ ounces saltpeter
1 pound brown sugar
4 quarts kosher-style or sea salt
3 ounces black pepper

1. Mix thoroughly in large bowl or pan the brown sugar, saltpeter and one quart of salt.

2. Rub thoroughly into hams, adding an extra amount of the mix to expose meat at butt and shank ends.

3. Let stand 24 hours, then rub meat with pepper and another quart of salt.

4. Let stand for nine days, then rub again with the remaining 2 quarts of salt.

5. Set aside for thirty days, then brush off excess salt.

6. Place each ham in a clean linen or muslin bag, nestled in chopped straw; tie neck of bag and hang in cool, dry, dim place until ready to smoke or cook as is.

7. Hams to be prime country cured should hang one year or longer, although they may be smoked after setting aside for the thirty-day period.

Dry Salting of Other Pork Cuts: The same method used to cure country hams may be used to dry-salt other pork cuts, such as roasts, chops, spare ribs, knuckles, feet and pork fat. Larger cuts over 10 pounds should be hung individually in bags; smaller cuts under 10 pounds may be hung together in the bags, but cuts should be kept from touching by laying thick layers of clean, chopped straw between each individual piece of meat. To prepare pork for cooking, soak in cold water for three days, changing water several times a day, then prepare according to your favorite recipe.

COUNTRY CURED BACON

100 pounds sides of bacon
8½ pounds salt
3¼ pounds brown sugar
3 ounces saltpeter
4 gallons water

1. Lay sides of bacon flat on wooden tabletop or large board and rub thoroughly with salt. Let stand, uncovered, for 48 hours.

2. Mix well saltpeter and brown sugar; dissolve in the water.

3. Heat water to boiling and cook for 20 minutes. Skim liquid and cool.

4. Place bacon in layers in clean oak barrel and pour liquid over meat. Place stone or other heavy weight on bacon in order to keep it under brine.

5. Let stand five weeks, then hang to dry before smoking. Bacon may be kept in brine for one year without spoiling.

COUNTRY CURED SAUSAGE

30 pounds pork
15 pounds clear fat pork
3 teaspoons sugar
1½ tablespoons sage
1½ teaspoons ginger
¾ pounds salt
3 tablespoons black pepper

1. Slice meat into 2-inch pieces and add the sugar and seasonings.

2. Feed through meat grinder, grinding twice.

3. Pack tightly into sterilized pint or quart jars, cover with ½ inch melted pork fat, cool, seal jars and keep in springhouse or other cold place.

4. To cook, shape into patties and fry in pan until golden brown.

HEADCHEESE

5 pounds head or shoulder pork
7 pounds side pork
Salt
Pepper
4 onions, quartered
2 carrots, halved
4 bay leaves
Ground cloves

1. Use separate kettles. Do not remove rind (skin) from meat.

2. In one kettle place fat-side meat and cover with water. In other, place lean meat from head; cover with water.

3. Divide onions, carrots and bay leaves between kettles.

4. Boil until meat is very tender, skimming fat off as it rises to surface.

5. While meat is still hot, prepare headcheese. Spread large piece of muslin in large, shallow pan.

6. Peel rind from meat and place as single layer on muslin, fitting pieces together, if necessary, to cover bottom. Alternate layers of fat and lean pork, seasoning each layer with salt, pepper and ground cloves.

7. Cover top and sides with more rind; tie up tightly in muslin to form bag.

8. Reheat in liquid, remove meat from muslin bag and press in flat dish under heavy weight.

9. Strain cooking liquid, cool until lukewarm, then pour over meat until just covered.

10. Store in cool place until ready to use.

Smoked Pork Select hams or other pork products, such as front shoulder (picnic) hams, bacon, knuckles. Remove from bags and hang from rack in smokehouse (see next chapter). Start small fire in fire pit, using dry kindling wood. When fire has burned down to red coals covered with film of ash, add two handfuls of hickory chips, apple wood chips or roughly ground corn cobs or an equal mixture of all three to the coals to produce a dense, aromatic smoke. Smoke for ten days, adding more chips as needed.

SMOKED COUNTRY SAUSAGE

30 pounds pork
15 pounds clear fat pork
½ ounce red pepper
4 tablespoons sage
¾ pounds salt
1¾ ounces saltpeter
3 tablespoons black pepper

1. Slice meat into 2-inch pieces and add the seasonings.

2. Feed through meat grinder, grinding twice.

3. Stuff sausage casings with mixture, tie ends securely and smoke as above for three to five days.

4. Hang in dry, cool place.

Meat in the Crock

Use this method for pork, veal, beef and lamb. Meat may be put down in stone crocks or large oak casks. If crocks are used, wash thoroughly and place upside down in pan of boiling water on top of stove. Boil 20 minutes, then remove without drying.

Roasts Wipe meat with damp cloth, season with salt and pepper to taste, and place in roasting pan. Roast in preheated, 350° F. oven for 35 to 40 minutes for center cuts, 40 to 50 minutes for rib or shoulder cuts. Pack in crocks or oak casks and cover immediately with hot fat. Or use your own favorite recipe, and continue as above. Store in cool place.

Chops and Steaks Season meat and sear quickly on both sides. Add 3 tablespoons of water. Cover tightly and cook over low heat. Cook single-cut chops and steaks 20 to 25 minutes; double-cut, 35 to 40 minutes. Pack closely in crocks or oak casks and cover immediately with hot fat. Store in cool place.

To Use Remove meat from crock or cask, scrape and wipe off as much fat as possible. Heat thoroughly in 350° F. oven. Insert meat thermometer and cook until done.

PICKLED TONGUE

½ teaspoon saltpeter
2 quarts water
1 cup salt
3 tablespoons sugar
3 cloves garlic
2 bay leaves
¾ teaspoon mixed spices
1 beef tongue

1. Wash tongue thoroughly. Rub with salt, sugar, pepper and other seasonings.

2. Place in crock or oak cask with all but one cup of the water.

3. Warm this one cup of water and dissolve saltpeter in it. Add to crock or cask.

4. Cover with large plate and weigh down with heavy stone to keep it submerged.

5. Store in cool place, turning once a week.

6. Leave in brine for three to four weeks, adding another ½ cup of salt after first week.

Corned Beef Use as much fresh-killed beef as desired. For large quantities, use 55-gallon oak barrels; for smaller quantities, use stone crocks or oak casks. Cover with cold water. Let stand two days. Drain off water and measure before discarding. Fill large container with same amount of fresh water. For every 100 pounds of beef used, add 2 pounds of salt, one pound of brown sugar and 2 ounces of saltpeter to container and boil 15 minutes on top of stove. Skim scum from surface, cool and pour over beef in barrel. Place heavy weight on beef to keep it submerged. Store in cool place. Beef will be ready for use in ten days, but can be kept in brine indefinitely.

DRIED BEEF

Fresh-killed beef
Salt
Saltpeter
Brown sugar

1. For every 100 pounds of freshly-killed beef use ½ pound of salt, 1½ teaspoons of saltpeter and 1/8 pound of brown sugar.

2. Mix seasonings well and divide into three portions.

3. Rub one portion thoroughly into meat and place beef in barrel or stone crocks. Let stand 24 hours. Repeat process every 24 hours for two days, turning meat three times a day.

4. After third day, allow meat, without further turning, to rest for seven days.

5. Hang in warm place until beef stops dripping.

6. When dripping stops, hang meat in cool shed for seven weeks or until thoroughly dry.

7. Wrap in clean muslin bags and store in cool place.

8. Every six months, when beef hardens, remove from bags and soak in cold water for 36 hours. Wipe dry and return to cool place in muslin bags.

Natural Freezing

Meat to be frozen should be hung outside in cold weather, high enough off floor of covered shed or porch to protect it from predatory animals. In the case of sides of hogs, beef, veal, etc., wrap well in muslin and tie securely. Fresh hams and very large roasts should be wrapped in freezer paper and placed, individually, in muslin bags. Smaller cuts should be wrapped in freezing paper and placed together in muslin or burlap bags with chopped straw to prevent the packages from touching each other. Meat will keep as long as cold weather holds.

A two- or three-day thaw will not spoil the meat, although it is best to remove it to a basement, root cellar or other cool place until the weather turns cold again. If thaw persists, the meat should then be dry-salted as described earlier in this chapter. To preserve meat in a home freezer, follow the instructions that come with the unit.

PREPARING BEEF PRODUCTS

Pickling Tripe

After you have thoroughly cleaned and rinsed the tripe in cold water, scald it in hot water (a little below the boiling point). When sufficiently scalded, remove the inside lining of the stomachs by scraping, which will leave a clean, white surface. Boil tripe until tender (usually about 3 hours) and then place in cold water so that you may scrape the fat from the outside. When you have done this, peel off the membrane from the outside of the stomach, and the clean, white tripe is ready for pickling.

Place the tripe in a clean, hardwood barrel or earthenware jar, and keep submerged in a strong brine for three or four days. Rinse with cold water and cover with pure cider vinegar or a spiced pickling liquid. Place a weight on the tripe to keep it from floating on the surface of the liquid.

Making Hamburger

Grind lean beef—such as the round, neck, flank and trimmings—with a little fat in a sausage grinder. If you desire, add a small amount of bacon for flavor. For seasoning, about one pound of salt and 4 ounces of pepper are sufficient for 50 pounds of meat.

Making Bologna-Style Sausage

In making bologna, for each 20 pounds of beef, add 5 pounds of fresh pork. Grind the meat coarse, then add the seasoning and grind through the fine plate.

For seasoning 25 pounds of meat, ½ pound of salt and 2½ ounces of pepper are usually satisfactory. Garlic may be added if desired.

Add 3 to 4 pints of water to this quantity of meat. Mix with the hands until the water is entirely absorbed by the meat. When thoroughly mixed, stuff into soaked beef casings or "rounds," and smoke the bologna from two to three hours at a temperature of 60° to 70° F.

After smoking, cook the bologna in water about 200° F., or slightly below the boiling point, until it floats.

Keep the sausage in a dry place for immediate use or pack it in cans; cover to within ½ inch of top with the liquid in which the bologna was cooked. Then heat it to a temperature of 250° F. for 45 minutes, or at 15 pounds steam pressure.

17

Preserving Fish and Shellfish

Since not all fish and shellfish are available at all seasons, it is a good idea to learn how to preserve them for future use. In the coastal and lake regions of this country and Canada, where fish and shellfish are considered an important supplement to the family diet, fish are frozen, canned, smoked, dry-salted and pickled.

All of the methods named have their advantages and disadvantages. Preserving fish in the home freezer is certainly the simplest method. Canning consumes more time, and requires considerably more equipment than any of the above methods, but most authorities believe that the product will keep longer. In many areas of the country, smoking and salting are used almost exclusively, because these methods are cheaper, simpler and require very little equipment.

Whatever method you decide to use, the first thing you must do is learn how to dress fish correctly. Fish must be kept cold from the minute they come out of the water until the time they are processed, regardless of what method you use. If ice is not available for cooling, gut the fish immediately and sprinkle the cavity with salt.

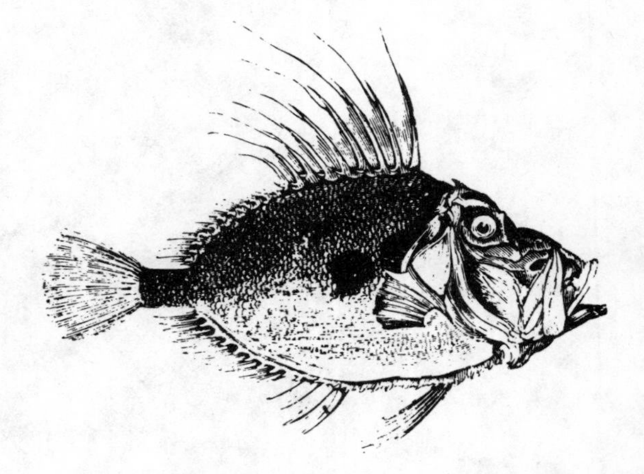

Even if cleaned immediately after being caught, fish must be further cleaned and washed before preserving can begin. Remove the fins, and scrape the fish until it is free of blood and slime, then cut off the head and tail and clean all remaining viscera or membranes from the belly cavity.

To make scaling easier, dip the fish quickly into boiling water. It will loosen the scales and also help to remove the slime. If scaling of fish is not done with the use of boiling water, remove the slime by washing in a vinegar-water solution of one part vinegar to four parts water, then rinse well in clean water.

Protecting Home-Cured Fish

Fish products are preserved to prevent spoilage during storage. Fresh, dried, salted or smoked fish are liable to attack by bacteria. Spoilage will take place fastest at temperatures of 70° to 100° F., so it is very important to keep the preserved product in a cool, dry and dark storage area.

Home-cured fish products should be protected against spoilage by being placed in tightly closed containers. Keep brine-cured products weighted down below the surface of the brine. Cover smoked products with a thin coating of paraffin, or wrap in oiled or parchment paper and place in a tightly closed container.

Preliminary Instructions for Canning

If you are going to can fish, the most essential piece of equipment you will need is a good pressure cooker. Pressure is absolutely necessary to produce a temperature above boiling (212° F.) that can be depended upon to kill the spores of botulinus, a highly dangerous food poison that has been found to have fatal results in three out of every five cases. Ten pounds of pressure, registering on the gauge as 240° F., will destroy these spores and remove any danger of contamination.

How to Use the Pressure Canner

Pour 3 inches of water in the bottom of the canner, then set the filled glass jars on the rack. Be careful not to let the jars touch each other or the sides or bottom of the canner. This goes for glass jars only. Tin cans may be packed in the canner close together, but not so close that the steam cannot flow around each can. When the canner has been loaded fasten the cover tightly so no steam can escape except through the open pet cock. Keep a close eye on the canner until the steam flows steadily from the opening. Allow it to escape for 10 or more minutes until all the air is driven from the canner. Then close the pet cock or put on the weighted gauge, according to canner directions. The moment the proper temperature is achieved, start counting for the time given in the directions for canning the product you are working with. To maintain the proper pressure, which must be kept constant, regulate heat under the canner. When the time is up, remove the canner from the heat at once.

Glass Jars

After removing the canner from the heat, wait until the pressure goes down to zero. After another minute, slowly open the pet cock or remove the weighted gauge. Unfasten the cover and tilt the side away from you *up* so that the escaping steam will not flow in your direction. Remove the jars and place them on a cloth-covered table, making sure they are protected from drafts. If you see that liquid in any of the jars has boiled away, do not open them to add more. Simply set them aside, and plan to use their contents first. Cool the jars right side up, label them with the date processed and the type of product they contain. Store in a cool, dark place.

MARKET FORMS OF FISH

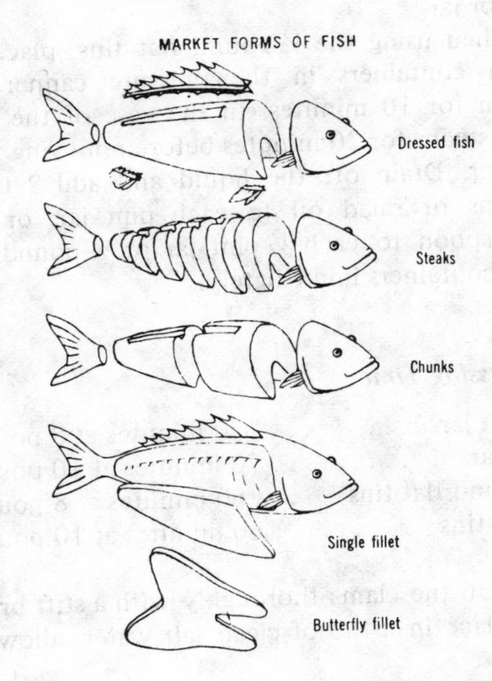

Tin Cans

Immediately release steam in canner when the processing time is up. Open the pet cock or remove the weighted gauge. When pressure reaches zero, wait one minute, unfasten the canner cover and tilt it away from you until the steam escapes. Remove the tins and cool in cold, clear water, changing the water as it becomes warm. Remove the cans from cold water while still lukewarm and set them aside to dry. Label as with jars, and store in a cool, dry place.

Salt Water Fish

Processing Time

½ pint jars . . . . 85 minutes at 10 pounds
Pint jars . . . . 95 minutes at 10 pounds
½-pound flat tins . . . . 75 minutes at 10 pounds
No. 2 tins . . . . 90 minutes at 10 pounds

Fillet the fish, making sure you remove the dark, fatty portions. Cut the fillets into can- or jar-size pieces. Make a brine by adding 1 1/3 cups of salt to one gallon of water. Add the cut pieces of fish to the brine and soak for 15 minutes. Remove and drain for 5 minutes, then pack tightly in containers, leaving ¼- to ½-inch headroom between the fish and the rim of the can or jar.

When using the ½-pound flat tins, place the open containers in the pressure canner and steam for 10 minutes. In the case of the glass jars, steam for 20 minutes before removing from canner. Drain off the liquid and add 2 tablespoons of salad oil to each pint jar, or one tablespoon to each ½-pint jar or ½-pound tin. Seal containers immediately.

Clams

Processing Time

½-pint jars . . . . 70 minutes at 8 pounds
Pint jars . . . . 70 minutes at 10 pounds
½-pound flat tins . . . . . 60 minutes at 8 pounds
No. 2 tins . . . . 60 minutes at 10 pounds

Scrub the clams thoroughly with a stiff brush and place in a tub of clean salt water, allowing 1 1/3 cups of clams to one gallon of water. Add a few handfuls of cornmeal to the brine and let stand 12 to 24 hours. Open clams over a large pan, and save the liquid. If littleneck clams are used, snip off the necks.

Remove meat from the old brine and wash well in a new brine made up of 2 ounces of salt to 1 gallon of water. Remove from this brine and blanch clam meat for one minute in a solution of ½ teaspoon of citric acid crystals to one gallon of cold water. Drain well and pack whole into containers.

In glass jars, pack 1½ cups or 12 ounces of meat into pint jars and fill within ¼ inch of rim with hot, strained clam juice. In tins, pack 1½ cups of meat into No. 2 tins and fill with strained clam juice.

Crabs

Processing Time

½-pint jars . . . . 85 minutes at 10 pounds
Pint jars . . . . 95 minutes at 10 pounds
½-pint tins . . . . 75 minutes at 10 pounds
No. 2 tins . . . . 90 minutes at 10 pounds

Place live crabs in ice water for 5 minutes. Kill with a sharp-pointed knife. Remove the claws and legs, and remove the back shell by placing fingers into the leg holes and pulling apart. Remove the viscera and wash thoroughly under running water.

Use a large pot or kettle to which has been added ¼ cup of vinegar and one cup of salt to each gallon of water. Bring the water to boil and add the crabs, boiling for 25 minutes. Pick the meat out of the shells. Break the claw and leg shells with a mallet and peel off the broken pieces, taking care not to break the meat.

Keep the body and leg meat separate. Immerse the picked-over meat for 2 minutes in a brine made of one cup of salt to one gallon of water. Remove from the brine, then dip into mixture of one cup of vinegar to one gallon of water. Press the meat between a clean cloth to remove excess moisture.

For glass jars, place ¾ cup of crab meat into half-pint jars, and 1½ cups of meat into pint jars.

Line the jars with a layer of leg meat in the bottom and around the sides. Fill the center with the body meat. Process as instructed above.

Fill tin cans with body meat and top with a layer of leg meat. As an added protection for the meat, the tins may be lined with vegetable parchment paper before filling. Cinch the covers loosely and exhaust the cans for 10 minutes at boiling temperature. Seal tightly and process.

Trout and Perch
Processing Time

Pint jars .... 100 minutes at 10 pounds
No. 2 tins .... 90 minutes at 10 pounds

Scrape off the scales and remove the fins and heads. Clean and wash thoroughly in cold water. If the fish are large, they must be cut in slices, otherwise pack whole. Add salt to cover, but do not add water. For glass jars, seal tightly and process. For tin cans, crimp the lid loosely and steam 15 minutes. Seal.

Spiced Fish
Processing Time

½-pint jars .... 100 minutes at 10 pounds
No. 2 tins .... 90 minutes at 10 pounds

Use salmon, shad, trout, lake trout or mackerel. Scrape off the scales and remove the fins and heads. Clean and wash the fish thoroughly in cold water. Leave the backbone intact. Do not remove or fillet. Cut the fish into pieces and soak in a brine made up of ½ pound of salt to one gallon of water for one hour. Drain the fish for 15 minutes.

VINEGAR SAUCE

2 quarts water
2 quarts vinegar
¼ cup sugar
¼ ounce whole white pepper
¼ ounce whole cloves
1/8 ounce cardamom seed, cracked
¼ ounce mustard seeds
1/8 ounce cracked whole ginger

1. Add sugar and water to the vinegar.
2. Tie the spices in a cheesecloth bag and simmer in the vinegar solution for 1½ hours.
3. Strain.
4. Fill containers with fish. Add the vinegar solution.
5. Place containers in kettle, but do not cover jars or tins. Cook for 20 minutes.
6. Invert containers on a wire screen and drain for 5 minutes.
7. Reverse containers and add a few slices of raw onion, a pinch of mixed spices, and one bay leaf. Add vinegar to cover fish.
8. Seal containers and process.

Smoking fish helps to preserve the meat six weeks or more, according to the method used. Smoking also adds a delicious flavor to the delicate flesh.

There are two methods of smoking fish: cold smoking, which will keep the fish without spoiling for six weeks or longer; hot smoking, which cooks the fish while they smoke, but will preserve them for no longer than two to three weeks without refrigeration.

Cold Smoking

Use any type of fresh-water or salt-water fish. Bleed and clean fish thoroughly immediately after they are caught. Keep them out of direct sunlight and handle them gently. Rough handling causes fish to spoil more readily.

Split fish down the back with a sharp knife, being careful not to cut all the way through at the tail end. If the catch is Spanish mackerel or equally fat fish, remove the backbone. The backbone is not removed from lean fish. It is left in to help retain their shape. Remove all traces of blood, using a stiff wire brush if necessary. Use a coarse cloth to remove the dark membrane in the belly cavity. Clean the fish thoroughly in cold water.

Make a light brine of 2 pounds of salt to 5 gallons of water. Immerse the fish for 30 minutes, then remove from brine and drain for 20 minutes.

Fill a shallow box or pan with a 4-inch layer of salt. Dredge the fish one at a time, rubbing the salt well into the skin and belly cavity. After removing the fish from the salt box, pack skin down in a tub or box with a thin layer of salt in the bottom. Place the fish in a layer at right angles to the one below it, scattering a handful of salt between each layer. Use about one pound of salt to each 3 pounds of fish. Place the top layer of fish skin side up.

Leave the fish in the salt for 3 hours, then rinse, being careful to remove all traces of the salt. Dry on wire racks for 3 or 4 hours or until a thin film appears on the surface. Place the fish in a smokehouse, stringing them on wooden rods. The rods are put through the fish just below the bony plate in back of the head, one in each side. A pair of rods 3 feet long will hold twelve or more fish. Space the fish at least one inch apart or the smoke will not penetrate properly.

The fire should be started in the fire pit 2 or 3 hours before the fish are hung in the smoke-

Drying rack

(Below) *Simple smokehouse*

house. Start the fire with two or three sticks of hardwood, approximately 3 feet long by 4 inches thick. Any nonresinous hardwood, such as hickory, birch, oak, alder beech or cypress will do. The fire must be kept low and smoldering. It should not be allowed to give off great clouds of smoke during the first 8 to 10 hours of smoking, although dense smoke *is* needed for the remainder of the smoking process.

Never allow the fire to blaze up, nor should the air in the smokehouse feel warm to the hand. Until the smoking process is completed, the fire must be tended during the night as well as the day. It must never be allowed to die out.

If the fish are meant to be kept six weeks or longer, they should be smoked for five days. A smoking period of 24 hours will keep fish from spoiling for two weeks or slightly longer. After smoking, remove the fish from the smokehouse and dry in the air for 2 to 3 hours. Sprinkle lightly with salt, wrap in waxed paper and pack in wooden boxes. Store in a cool dry place.

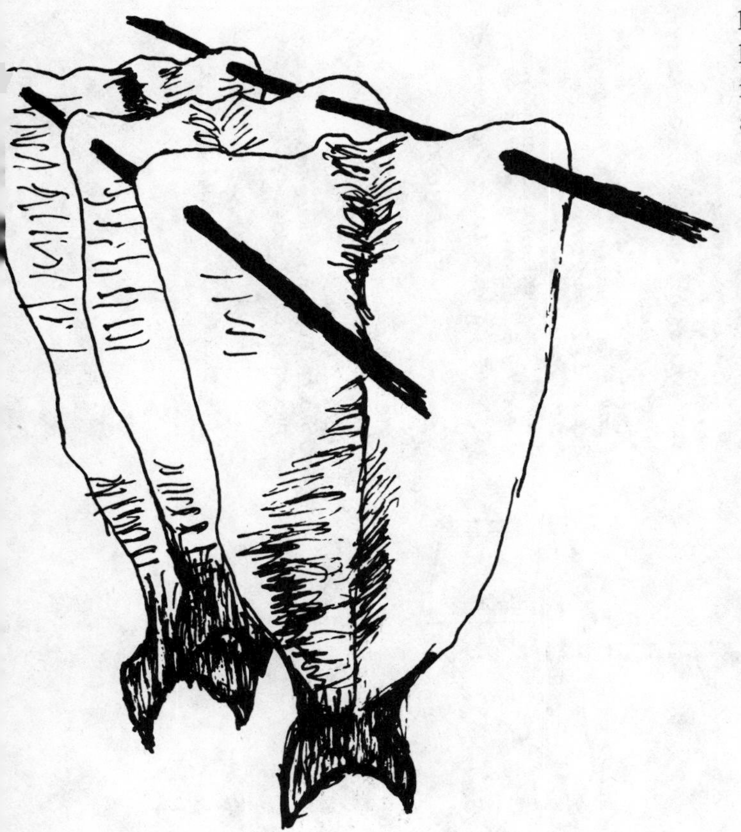

Fish strung on wooden rods

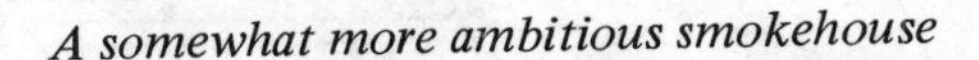

A somewhat more ambitious smokehouse

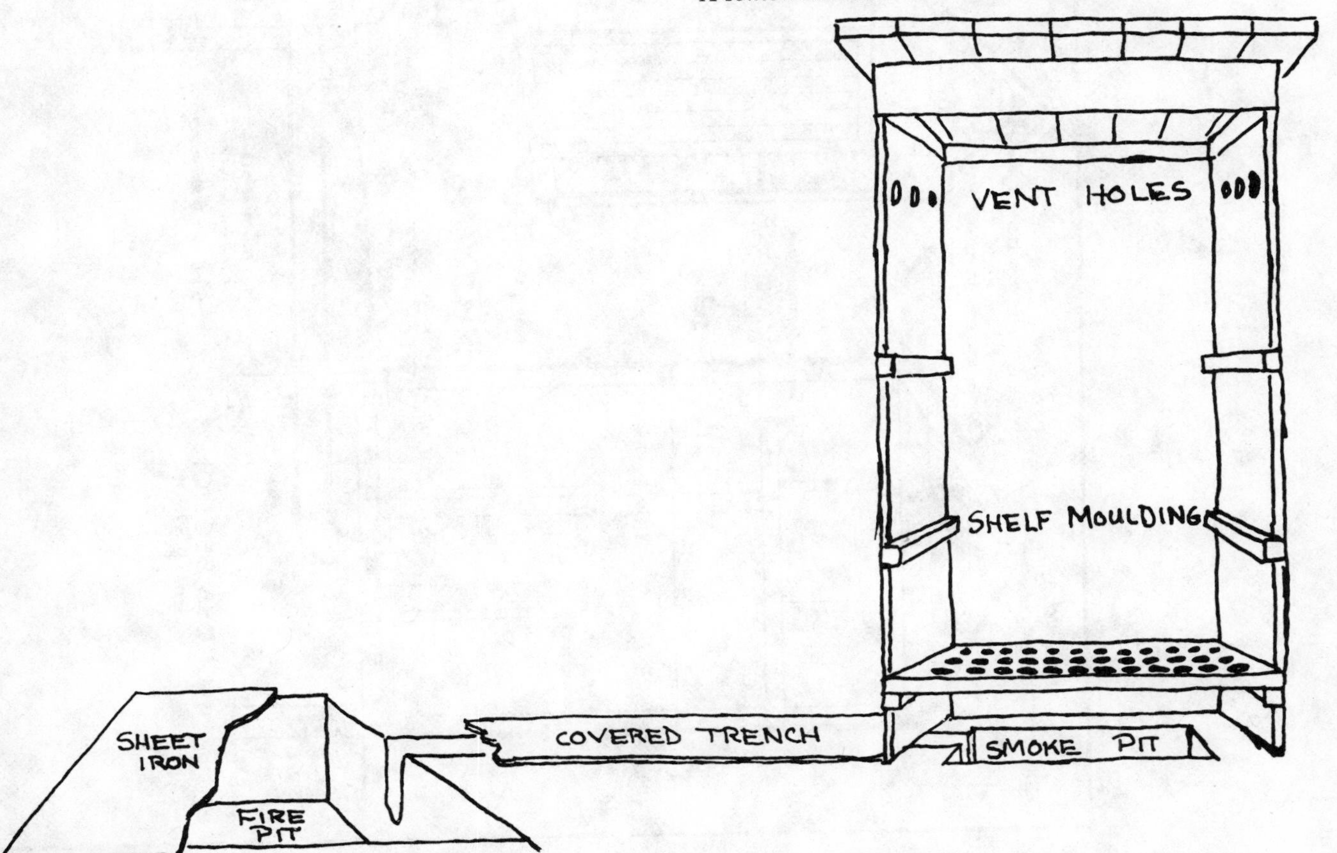

Detailed plans for building a good, permanent smokehouse

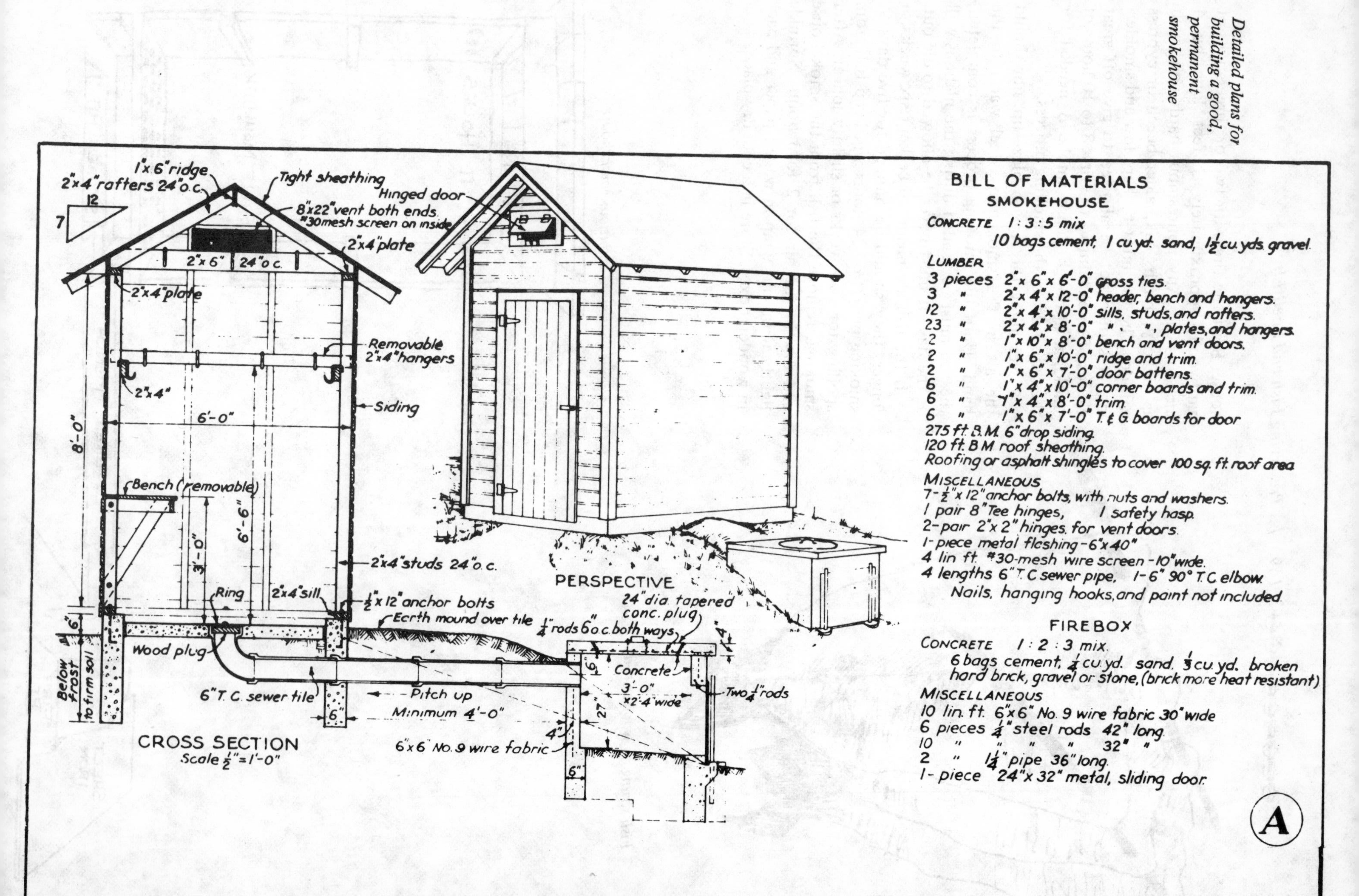

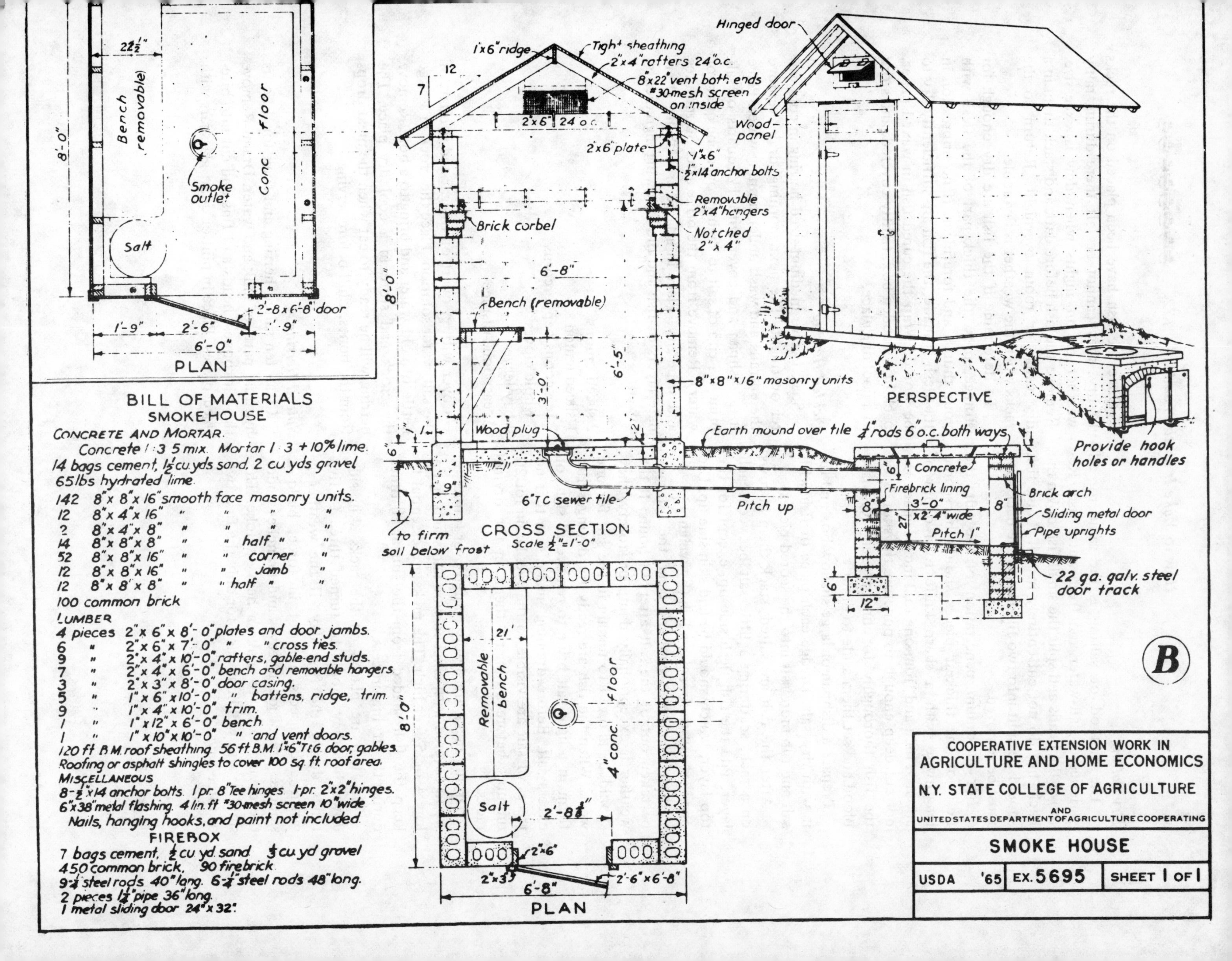

PLAN
Bench (removable)
Smoke outlet
Conc floor
Salt
2'-8 x 6'-8 door
8'-0"
22 1/2"
1'-9"
2'-6"
1'-9"
6'-0"
BILL OF MATERIALS
SMOKEHOUSE
CONCRETE AND MORTAR.
Concrete 1:3 5 mix. Mortar 1:3 + 10% lime.
14 bags cement, 1 1/2 cu. yds sand, 2 cu yds gravel
65 lbs hydrated lime.
142 8" x 8" x 16" smooth face masonry units.
12 8" x 4" x 16" " " " "
2 8" x 4" x 8" " " " "
14 8" x 8" x 8" " " half " "
52 8" x 8" x 16" " " corner "
12 8" x 8" x 16" " " jamb "
12 8" x 8" x 8" " " half " "
100 common brick
LUMBER
4 pieces 2" x 6" x 8'-0" plates and door jambs.
6 " 2" x 6" x 7'-0" " " cross ties.
9 " 2" x 4" x 10'-0" rafters, gable-end studs.
7 " 2" x 4" x 6'-0" bench and removable hangers.
3 " 2" x 3" x 8'-0" door casing.
5 " 1" x 6" x 10'-0" " battens, ridge, trim.
9 " 1" x 4" x 10'-0" trim.
1 " 1" x 12" x 6'-0" bench.
1 " 1" x 10" x 10'-0" " and vent doors.
120 ft B.M. roof sheathing. 56 ft. B.M. 1"x6" T&G. door, gables.
Roofing or asphalt shingles to cover 100 sq. ft. roof area.
MISCELLANEOUS
8-1/2" x 14 anchor bolts. 1 pr. 8" Tee hinges. 1 pr. 2"x2" hinges.
6" x 38" metal flashing. 4 lin. ft #30-mesh screen 10" wide
Nails, hanging hooks, and paint not included
FIREBOX
7 bags cement, 1/2 cu yd. sand 1/3 cu. yd gravel
450 common brick, 90 firebrick
9-1/4" steel rods 40" long. 6-1/4" steel rods 48" long.
2 pieces 1 1/2" pipe 36" long.
1 metal sliding door 24" x 32".
1"x6" ridge
Tight sheathing
2"x4" rafters 24" o.c.
8"x22" vent both ends #30-mesh screen on inside
2"x6" 24" o.c.
2"x6" plate
1"x6"
1/2"x14" anchor bolts
Removable 2"x4" hangers
Brick corbel
Notched 2" x 4"
6'-8"
8'-0"
Bench (removable)
6'-5"
3'-0"
8"x8"x16" masonry units
Wood plug
Earth mound over tile
1/4" rods 6" o.c. both ways
6" T C sewer tile
Pitch up
to firm soil below frost
CROSS SECTION
Scale 1/2" = 1'-0"
Concrete
Firebrick lining
Brick arch
3'-0" x 2'-4" wide
Sliding metal door
Pipe uprights
Pitch 1"
22 ga. galv. steel door track
12"
Hinged door
Wood-panel
PERSPECTIVE
Provide hook holes or handles
Removable bench
4" conc floor
Salt
2'-8 3/8"
2"x6"
2"x3"
2'-6" x 6'-8"
6'-8"
21'
PLAN
B
COOPERATIVE EXTENSION WORK IN
AGRICULTURE AND HOME ECONOMICS
N.Y. STATE COLLEGE OF AGRICULTURE
AND
UNITED STATES DEPARTMENT OF AGRICULTURE COOPERATING
SMOKE HOUSE
USDA '65 EX. 5695 SHEET 1 OF 1

Dry Salting

The method of salting is the same for all varieties of salt-water and fresh-water fish. Remove the heads and split the small fish down the back, but do not cut entirely in half. Large fish should be split into two fillets and the backbone must be removed.

For maximum penetration of the salt, the flesh of the thickest pieces should be scored lengthwise with a sharp knife to a depth of about ½ inch and from one to 2 inches apart. Do not cut deep enough to puncture the skin. Wash the fish thoroughly to eliminate any diffused blood and set them aside to drain.

Prepare a dishpan or large shallow box and fill it with dry salt. Dredge each piece of fish with salt and rub more salt into the scored pieces.

After the fish are salted, stack them in rows on a rack, scattering a little salt between each layer. Pile the fish flesh side up, except for the top layer, which should be laid skin side up. Use one part salt to four parts fish (by weight).

If the weather is cool and clear, the fish may be taken out of the salt after 48 hours. If the weather is exceptionally damp or stormy, they should be allowed to remain in the salt for one week. When the fish are ready for drying, scrub them in a light salt brine to remove all signs of excess salt. Before draining, make sure that no traces of salt are visible to the eye. Drain for 20 minutes before arranging them on the drying racks.

Drying racks are simply roofed over frames of wood covered with chicken wire and standing on legs 4 or 5 feet high. The racks are essential to keep the fish shaded from the sun at all times during the drying process.

The fish are placed on the racks skin side down but they should be turned three to five times during the first day. If the weather is warm and insects are out in force, build a smudge fire of green leaves and branches under the drying racks to make a heavy smoke. This smudge fire will be needed for only the first two or three days.

Once the fish have been placed on the racks to dry, they cannot be left there during rainy weather or on nights when dew is excessive. Remove the fish before dark and stack them in a dry cellar or room overnight. Return to the racks when the weather is favorable.

To determine if the fish are dry enough for storage, press the thick part of the flesh with your thumb and forefinger. If no impression in the flesh is made, the fish are sufficiently dry to be stored. Wrap the cured fish in waxed paper, pack in a wooden box, tightly covered, and store in a cool dry place.

Pickled Herring

Soak twelve medium-sized herring in cold water overnight. The next morning, drain, remove the entrails and wash under running water. Place in a dishpan and cover with 1½ quarts of salt brine (½ cup of salt to slightly over one quart of water). Remove from the brine after 30 minutes and wipe dry. Place in a large crock with alternate layers of fish and the following ingredients, mixed:

2 sliced onions
1 sliced lemon
2 teaspoons peppercorns
8 whole cloves
4 bay leaves
½ teaspoon Allspice

Cover with ¾ quart of vinegar and ¾ quart of the salt water solution. Place the crock in a large kettle of cold water and bring to a boil. Remove the crock and store in a cool dry place. The herring will be ready to eat when the meat drops from the bones. Chill before serving.

Spiced Herring

Clean ten salted herring and soak for 4 hours in equal parts of milk and water. Drain. Remove the backbones, heads and fins and cut into bite-sized pieces. Place in pint or quart glass jars with alternate layers of sliced onions. Set aside while making the following mixture:

2 cups vinegar
2/3 cup salad oil
2/3 cup water
½ cup of sugar
8 bay leaves
1½ teaspoons peppercorns

Combine all ingredients and bring to a boil. Cool, pour over herring in the jars, and seal. The herring will be ready for use after 48 hours. If stored in a cool dry place, they will keep for six weeks or longer.

18

Fish and Wild Game Management

Any reader who has stayed with me this long will have long since learned at least one thing from this book: A farm—small or large—is not simply a piece of land with a few buildings, a vegetable garden, some domesticated livestock and a patch of woodland. It is a complex living community based on the soil.

Working with the soil, and dependent on it, are plants and animals that convert plant nutrients, moisture and sunshine into food and fiber for man. A farm is a successful community only if all the living things in it are working for the benefit of the whole.

If there are not enough grasses in the community, the soil loses its ability to take up and hold moisture. It may become eroded and lose its power to produce corn for hogs. If there are not enough earthworms, the end result will again be soil erosion and community defeat. If there are not enough squirrels, there will be fewer acorns planted, resulting in fewer oaks. If there are not enough rabbits, there will be fewer foxes to feed upon them, and those that remain will look to the farmer's poultry for food.

On the other hand, if there are too few foxes to eat meadow mice, there will be so many meadow mice that there will be less alfalfa for dairy cows. If there are not enough songbirds, there will be too many destructive insects, resulting in a shortage of grain for beef cattle. But there must be useful insects to pollinate alfalfa, red clover, and sweet clover, or else these legumes will not produce seed.

So we see that a successful farm community must have an abundance of useful wildlife and a low number of the harmful kinds. It must have what is called a *favorable biologic balance.*

A system of farming that supports a family well without depleting basic resources has a favorable biologic balance. Conversely, farming (commercial-scale farming as practiced in the Midwest is a good example) that depletes the soil and results in plagues of insects, weeds and crop diseases has an unfavorable biologic balance.

You may be surprised at the wildlife population living on well-managed farms. Studies made on farms having soil conservation plans in effect tell the story. On a 100-acre farm with one-third of the fence rows in woody cover, 15 acres in protected woods, 25 acres in good pasture and 60 acres in a four-year rotation with two years

Woodchuck

of meadow and two years of crops, the useful wildlife population was estimated to be: (1) several million beneficial insects, mostly destroyers of other insects and some that help to pollinate fruits and legumes; (2) more than four hundred beneficial birds, of forty kinds; and (3) more than one thousand beneficial small mammals, many of which are effective insect destroyers.

Also present on well-managed farms are the game birds and animals—quail, pheasants, grouse, rabbits, squirrels, and on some farms, ducks and deer—that offer sport and food for you and your friends.

Fur-bearing animals such as mink, muskrats, raccoons, skunks and opossums can provide recreation and cash income for you and your family. These valuable animals occur in greatest numbers on farms that use the land wisely and provide places for them to live.

Colorful, energetic songbirds add much to the enjoyment of rural life through their music and their movements. Who doesn't enjoy the songs of the mockingbird and the meadowlark—or the sight of robins busily feeding their young?

Bumblebees, leaf-cutting bees, syrphus flies and other wild insects, formerly much more abundant than they are now, once helped farmers to produce legume-seed yields four times as high as those obtained today. Their numbers can be increased through good land management.

Large-mouth bass, bluegills, channel catfish, and sometimes trout, supply fun and food for farm families fortunate enough to have a farm pond. Many farm ponds produce from 150 to 250 pounds of fresh fish per acre each year.

(Above) *Muskrat*

(Below) *Raccoon*

Wildlife Requirements

While no two kinds of wildlife have exactly the same requirements for living, it is safe to say they all need food, cover and water.

To be really useful, food must be plentiful and close to cover that will furnish protection from enemies and weather. Also, it must be available in the seasons when food is scarce.

Autumn olive feeds game in the East.

On most farmland in the United States there is enough food from late spring to fall. Insects, wild fruits, weed seeds, waste grain, nuts or green plants are available. The critical season is winter. There are no insects. Many wild fruits are gone. Snow and ice may cover waste grain. Early spring is often as difficult as winter, with its cover of deep snow.

In the South, planting perennial food-producing plants close to good cover is the best way to be sure you have enough wildlife food throughout the year. In the North, you can extend cover plantings close to natural food sources or leave unharvested a part of the grain or other seed crops close to good cover.

Most kinds of wildlife need several kinds of cover. Cover must conceal nests and young, provide shade from the hot sun and shelter from chilling rains. It must allow escape from enemies and must protect against snow, sleet, cold and wind in winter.

Well-managed cover preserves contain these three essentials—unburned, ungrazed, unmowed grass for nesting; dense or thorny shrubs for protection from predators and for nesting; and clumps of evergreens for winter protection in the North. All three kinds of cover should be close together and close to available food supplies.

Wildlife obtain water from three sources: surface water, food that contains lots of water, and dew. In the East, upland wildlife can survive on succulent foods and dew. Surface water is a necessity for all wildlife in the arid West, as it is everywhere for water-loving species like ducks, muskrats and mink.

Only a few limiting factors can be controlled by man. The species factors are unchangeable, as is the weather. The effects of predators may be modified somewhat, but with uncertain results. Little can be done about diseases and parasites. Some of man's activities, such as time of plowing, could be changed; others, such as time of mowing meadows, cannot be altered much. Fortunately the greatest changes can be made in the most important factors. Success in managing land to produce useful wildlife lies in improving the amount, quality and distribution of food, cover and water.

Land primarily suited for use as cropland, pasture and woodland produces wildlife as a secondary crop. In addition, every farm has land that can and should be used to produce useful wildlife as a primary crop—it is wildlife land.

Wire fences offer no wildlife cover.

Blackberries are high in moisture.

Small areas well distributed over the farm, when coupled with proper use and management of the other land, make the whole farm an efficient unit for the production of all crops, including wildlife.

Cropland management practices helpful to useful wildlife include:

1. Crop rotations that include grass-legume meadow.
2. Liming and fertilizing.
3. Strip cropping.
4. Use of cover crops.
5. Stubble-mulch tillage.
6. Delaying mowing of watercourses and headlands until after grain harvest.
7. Spring plowing.
8. Leaving 1/8 to ¼ acre of grain standing next to good cover.
9. Spreading manure near cover in winter.

Harmful practices are burning, clean fall plowing, early mowing of watercourses and headlands and indiscriminate use of insecticides and weed killers.

Pastureland management practices helpful to useful wildlife include:

1. Grazing kept within the carrying capacity of the pasture.
2. Liming and fertilizing.
3. Reseeding or renovating.

Harmful practices are burning, too-heavy grazing, and complete clean mowing early in the season.

Woodland management practices that encourage wildlife include:

1. Protection from fire and grazing.
2. Selective cutting in small woodlands.
3. Leaving two den trees per acre when cutting timber.
4. Piling brush near the edge of the woods.
5. Leaving fallen hollow logs.

6. Clear-cutting of only small areas in large woodlands.

Harmful practices are burning, excessive livestock grazing, clear-cutting of large areas and cutting out all den trees.

Wildlife land consists of usually small areas that cannot be used economically to produce other crops but that are well adapted to the production of useful wildlife. Eight kinds of wildlife land are especially important. They are drainage-ditch banks, fence rows and hedges, marshes, "odd areas," ponds and pond areas, shelterbelts and windbreaks, streambanks and wildlife borders. Their management is discussed on the following pages.

Drainage-Ditch Banks

Your local soil conservationist is the man who can give you the best advice on soil treatments, kinds of grasses to seed and seeding methods. He can also tell you what legumes to mix with the grass.

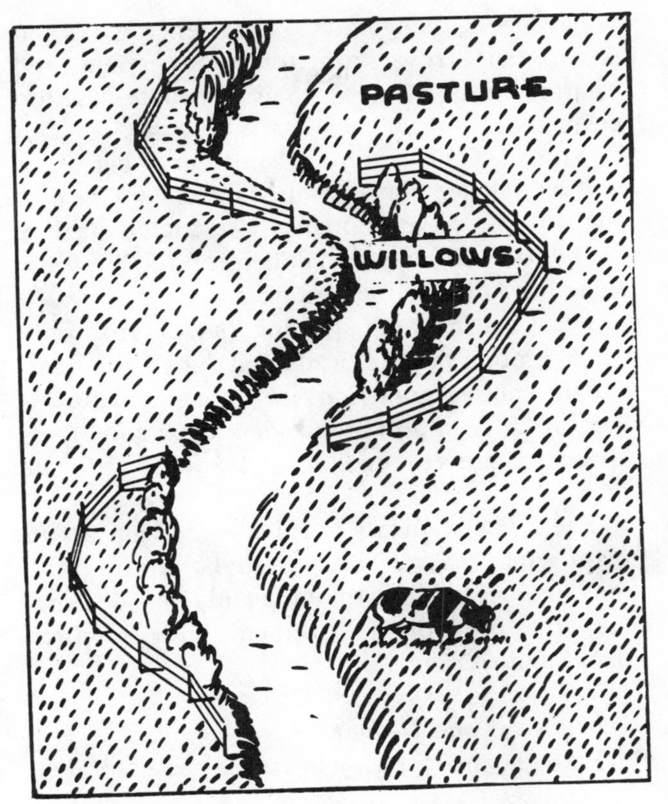

(Above) *How to protect from grazing streams having deep water and high banks*

(Below) *How to protect streams having shallow water and low banks*

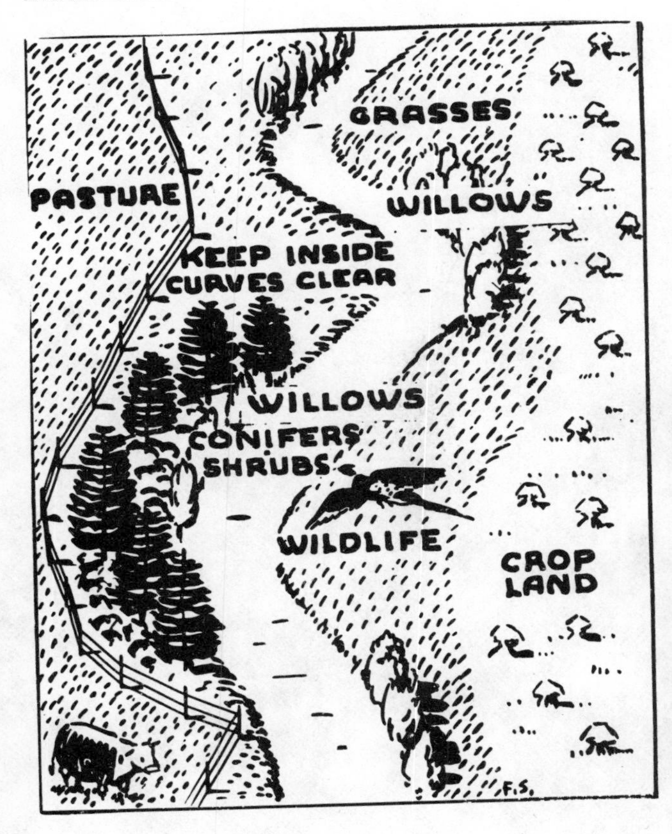

(Left) *Narrow strips of cover, like hedges and vegetated fence rows, are often called "travel lanes" because they serve to give wildlife ready access to all parts of a terrain well-supplied with them.*

Shrubs suitable for windbreaks include: bush honeysuckle, Persian lilac, autumn olive and Amur and California privets in the East; caragana, chokecherry, Russian olive, squawbush and wild plum in the West. Native lilac, climbing

Russian olive feeds Western wildlife.

honeysuckle, hedgerows, stone fence rows and stump fence rows from tree cuttings and land clearing furnish more than adequate wildlife cover the country over.

You can keep ditch banks and berms in grasses and legumes by mowing once a year if the slopes are flat enough, by restricted grazing, or by the use of weed killers of the type designed to kill broad-leaved plants.

Do your mowing, grazing or spraying only after ground-nesting birds have left the nest, usually about grain harvest time. Avoid overgrazing—it is important to maintain a good grass cover to prevent erosion and siltation of the ditch.

Keep woody plants *out*; keep good grasses *in*; plant windbreaks where needed; be careful to do mowing, grazing or spraying only in late summer—that is good management both for ditch banks and wildlife.

Fence Rows and Hedges

Barbed- and woven-wire fence, easy to erect but less wasteful of valuable cropland (for the large-scale commercial farmer), can be used successfully *if* native shrubs are allowed to grow in it.

Contour hedges save soil and game.

A new kind of fence is gaining popularity—the living fence of multiflora rose. It promises to be a real boon to wildlife.

At the same time many old ideas about the necessity for maintaining "clean" fences in order

Native shrubs improve wire fences.

Highbush cranberry; northern shrub

to control insect pests or weeds are going out the window. Modern studies of wildlife relationships are showing that woody fence rows have many advantages for the farmer.

Woody fence rows have been shown to harbor fewer harmful and more beneficial kinds of wildlife on general farms than do grassy fence rows. On farms producing small fruit and orchard crops, woody fence rows may be hazardous because they may harbor insects detrimental to those crops and may spread diseases common to woody plants. The danger is slight, however, if spraying programs are carried out. (Consult your County Farm Agent.)

Woody fence rows fit best where fence lines will not be changed, as between cropland and pasture, along property boundaries and streams, or around large gullies, ponds and odd areas.

Where multiflora rose cannot be grown, one of the following shrubs will produce good hedge or fence-row cover: red cedar, gray dogwood, American hazelnut, bayberry, silky cornel, highbush cranberry, bush honeysuckle, autumn olive, Russian olive, sand cherry, wild plum, trifoliate orange and squawbush.

Shrubs respond well to mulching. Use strawy manure, plain straw, old stack bottoms, sawdust, wood chips or stalks.

For fence rows in which you don't want to grow shrubs, you can make an improvement for wildlife with sweet clover. Simply plow a furrow to the fence row in the fall. In late winter or early spring broadcast (scatter) 30 pounds of sweet clover seed to the acre.

Existing fence rows composed of shrubs, trees and vines can be made neat in appearance, and the competition with crops reduced by cutting out the trees.

Marshland

On many farms there are areas of wet land. Some of them, particularly small potholes from ¼ to 5 acres in size located in the northern states, are valuable producers of waterfowl. Large areas of coastal marsh in the East and South are important wintering grounds for ducks and geese.

Many wet-land areas produce an annual catch of six to fifteen muskrats per acre. Also often found here are waterfowl, mink, raccoons, pheasants, prairie chickens and songbirds.

Muskrats should be trapped heavily—you should take sixty to seventy percent of the muskrat population annually—for cash income *and* to keep the population under control. Otherwise the colony will increase so rapidly that it will starve for lack of natural plant life.

For small marshes, the simplest management is prevention of grazing and uncontrolled burning. Marsh plants make better food for muskrats and waterfowl than for livestock. Grazing destroys valuable food and cover.

White pine "stump fence" used to protect fields from wandering livestock

Odd Areas

Small pieces of "waste" land that can be changed into wildlife land include small eroded areas in crop fields, bare knobs, sinkholes, small sand blowouts, large gullies, abandoned roads and railroad rights of way, borrow pits, gravel pits or even bits of good land that are cut off from the rest of a field by a stream, drainage ditch or gully. To be useful for wildlife, odd areas should be at least ¼ acre in size.

If you live in a pheasant-producing area west of the Mississippi River and want to increase pheasants, plant two rows of such hardy shrubs as wild plum, sand cherry, Russian olive or bush honeysuckle on the west and north sides of your farm boundary. Then sow a strip 100 feet wide to sweet clover. Your planting should be at least one acre in size to provide the winter feed and the nesting cover needed in the area.

Pond Building

Ponds are constructed for–

1. Soil erosion and flood control.
2. Water storage for livestock, supplemental irrigation, orchard spraying or fire protection for farm buildings.
3. *Fish production.*

In addition to fish, they can be expected to provide water and cover for fur-bearing animals, game birds and animals, and songbirds.

Ponds must be located on relatively tight subsoils or they will not hold water. They are built in small valleys with steep sides and gradually sloping floors so they will hold an adequate amount of water without excessive height in the fill. They should be placed where they will contribute to sound land use on the rest of the farm and where they will be free from excessive siltation.

The water supply for most ponds comes from rain water running off the land. The size of watershed needed varies according to local rainfall, topography, type of cover and rate of evaporation. The entire watershed should be in ungrazed woods or improved permanent pasture

or range. Cropland in the watershed shortens the life of the pond by allowing too much silt to get in. Springs are often suitable sources of water supply, but running streams should be avoided. They carry too much water and silt and make it difficult to keep stream fish out of the pond. (Consult your County Farm Agent for pond construction plans and for advice about local problems.)

On most farms, the pond should be fenced to keep out livestock. This is important to help prevent spreading livestock diseases; to protect the fill, spillway and pond edges from trampling; and to provide a filter strip to remove silt from the water before it reaches the pond. Fencing will also allow you to make plantings useful for upland wildlife.

All raw areas above the water line should be seeded to adapted grasses and shallow-rooted legumes such as alsike clover, red clover, Korean or common lespedeza. Never plant trees, shrubs or deep-rooted legumes like alfalfa or sweet clover, because roots of these plants will seek and find water in the pond, literally killing themselves and leaving tangles of dead trees, shrubs, etc. Use lime, fertilizer and manure as needed.

To manage your pond for fish production, don't plant anything in the water. Keep woody plants back at least 15 feet from the water's edge to allow room for fishing.

If you don't care about fish but want to attract muskrats and waterfowl to your pond, plant a few cattails, arrowhead, bulrushes, or burr reeds. These, or other native water plants, will eventually come in naturally if no planting is done.

For fish production your pond needs to be at least ¼ acre in surface area if you plan to fertilize systematically. If you don't plan to use a fertilizer your pond should have a surface area of at least ½ acre.

In the North, ponds should average 10 feet in depth with one-quarter of the pond 12 to 14 feet deep. In the Central States, ponds should average 6 feet in depth with one-quarter of the area 8 to 10 feet deep. In the South, they should be at least 6 feet deep in the deepest part.

Ponds for fish production should be built so they are 3 feet deep within 10 feet of the shoreline. Farm ponds may be stocked with fish as soon as the water approaches the spillway level, because then water is deep enough to support fish life.

In most states, fifty bass fingerlings and five hundred bluegill fingerlings per acre are stocked in unfertilized ponds. In fertilized ponds the stocking rate may be increased to one hundred bass and one thousand bluegill per surface acre.

An equal number of channel catfish may be substituted for one-quarter of the bass or stocked in addition to the bass and bluegills after the bluegills have become established. Channel catfish seldom spawn in farm ponds; they must be replaced occasionally.

In spring-fed ponds where the summer water temperature will not exceed 75° F., brook or rainbow trout may be stocked at the rate of three hundred to four hundred fish per surface acre. No other game fish should be stocked in trout ponds, but fathead minnows can be used as forage fish. Rainbow trout do not reproduce in ponds; they must be restocked periodically.

Bass, bluegill, channel catfish and trout fingerlings can be secured from hatcheries of the U.S. Fish and Wildlife Service, some state conservation agencies, or commercial hatcheries.

Most farm ponds are not fished heavily enough. Fishing for bluegills should begin the first summer after stocking. Bass should not be taken until they have reproduced; usually the first summer in the South but most likely the second summer in the North.

Note: Pond construction requires skill. Ask a neighbor or farm agent to recommend a contractor.

Fish farm ponds produce other sources of income. Large farm ponds attract wild migrating fowl which, in turn, can become an additional source of cash income from sportsmen who will gladly pay any reasonable fee for shooting rights. As for the fish in the pond, the same is true (during legal fishing seasons) of fishermen.

In the hands of a skilled fisherman, artificial flies (size 10 to 14) with light tackle will catch

This pond is in an area formerly used as a barnyard before being converted to an income-producing recreation enterprise.

many bluegills and provide real fishing enjoyment.

Bass are best caught with minnows or artificial bait. A "popping bug" on a fly line will give you a thrill when a bass takes it. Plugs or spinners used with a casting rod are also good. And, of course, minnows are the "old reliable." Bass fishing is best in spring, may be poor in summer.

Fertilizing farm ponds is often necessary for maximum fish production. A highly fertile pond will support more pounds of fish than one of low fertility. Experiments in the Southeast show that unfertile ponds support less than 100 pounds of fish per acre, average ponds support 150 pounds, highly fertile ponds support 400 pounds.

Fertile water shades out submerged weeds. It is greenish and opaque, because innumerable tiny plants are present. Light does not go any deeper into such water than you can see, of course, and weeds cannot grow without light. Although some people may object, it is perfectly safe to swim in fertile water.

Fertilizing costs money—$10 to $30 per acre per year. It will pay you dividends only if you fish the pond heavily or if you need to control water weeds.

Fertilizing must be done systematically. Good results cannot be obtained by haphazard methods. Once begun, it must be continued because the fish in the pond quickly put on weight. If fertilizing is stopped, the amount of fish food is sharply reduced and the fish must lose weight if they are to adjust to the lower amount of food present.

If you plan to use fertilizer, you should apply it at one- to two-week intervals beginning early in the spring. Continue to apply it until the water becomes so cloudy that you can't see the bottom in water that is 12 inches deep. Additional applications should be made until frost whenever the water begins to clear.

Each application of fertilizer should consist of 100 to 200 pounds per surface acre. You need nitrogen and phosphorus in about equal parts with one half to one-fourth as much potash. Desirable formulas are 8-8-2, 8-8-4, 10-10-5, or 12-12-6. If you cannot buy these, your soil conservationist, county agent or fertilizer dealer can tell you how to mix locally available fertilizers for a satisfactory formula.

Controlling of pond weeds is necessary for good fish production. Fertilizing is the most practical way to control underwater plants. Plants growing around the shore line are best removed by simply pulling them out as fast as they get started. Two or 3 hours a month spent at this task will usually provide satisfactory control.

Winter-killing of fish in northern ponds may occur when snow accumulates on top of the ice. The only practical recommendations for its control are to remove the snow from part of the pond surface or to lower the water a few inches below the ice to admit air.

Farm ponds can furnish a lot of pleasure—swimming, fishing, hunting, and in the North, skating. But they will do so only if they are well planned, well constructed and well managed.

Fish from a well-managed stock pond; red ear bream, bluegill bream and bass are those shown.

Fish Baits

Crayfish: Crayfish are crustaceans inhabiting stream riffles and shallow water areas of ponds and lakes; they are most often found under boards, stones and other objects. They may be caught with an 18 × 30 × 18-inch light wooden framework covered with galvanized screen wire. The large, Southern variety are too delectable for fish bait. Simmer in salted water until red and eat them.

Earthworms: Earthworms are the most common fish bait known to man and the most numerous. During humid, summer nights, collect them on the surface of well-fertilized sod fields or lawns, using a flashlight or lantern equipped with a red lens.

Crickets: Crickets, too common throughout the country for comment, may be raised in garbage cans, lard cans or metal drums with tops removed. Store them in a dry garage or basement and cover with a fine mesh screen as protection against natural enemies. First, sandpaper and coat the inside of the container with wax or varnish. Then cover the bottom of the container with 6 inches of fine sand. Add twenty to thirty adult crickets and feed with 2 pounds daily of poultry laying mash per hundred crickets produced.

Catalpa Worms: The catalpa worm is the caterpillar or sphinx moth in its larval stage. Since it can only be found on catalpa trees, its sole source of food, it must be harvested by shaking the tree and picking the worms from the ground. Catalpa worms may be kept alive in a cage and fed on catalpa leaves or kept in the refrigerator or springhouse in a container of cornmeal at least one gallon in size.

Frogs: Frogs make excellent bait for bass, pickerel, walleye and other fishes but should not be caught until they are either in the adult stage or have, at least, grown their legs. Catch them at the water's edge along stream banks, lake shores and swampy areas. They may be kept alive for long periods in a damp place, out of the sun, or in live bait boxes or sacks, in a pond or stream. Enough of the container should be above the surface of the water to allow frogs to surface for air.

Salamanders: Salamanders also make excellent fish bait. Collect them in springs, under logs and flat rocks along stream and lake shores. They may be kept alive several days if placed in wet moss and covered with wet burlap. Keep them damp in a cool place.

Collecting and Holding Fish for Fish Bait

Minnows: Bluntnose, fathead and golden shiner minnows are all to be found in ponds or sluggish streams spread over a large area of the United States and Canada, where they propagate at the rate of 100,000 to 200,000 per acre. They spawn under logs, stones and other objects at depths of approximately 18 inches. Propagation may be aided by adding sticks, boards and other objects to the pond or stream at about that depth, but if enough natural aquatic vegetation is present these aids are not needed. They feed upon small crustacea, microscopic plants and small aquatic insects.

Stream - Spawning Chubs: The creek chub, hornyhead chub, stoneroller, white sucker and chubsucker are all stream-spawning chubs. They spawn on gravel beds, stream riffles or on lake beaches where the wind action creates a strong current. Since propagation of chubs is a delicate task for the beginner, it is best to consult your local county or state hatchery for information.

Cut Baits: Fresh meat of the poorest grades (pork, beef, etc.) is often used by fishermen but with mixed results. Also used are garfish, mudfish, suckers and other nongame fish with far superior results. This writer has had the best results with the meat of rodents and other small pests, cut into small pieces and used only when fresh. The intestines or viscera of fowl and small mammals also serve as excellent fish bait.

Carp Bait: Carp bait is essentially small dough balls, firmly rolled so they will not fall apart in the water. Since every carp fisherman has his own favorite recipe, I am including only those I have used myself or have seen used successfully.

Carp Bait (1): ½ cup of plain flour, ½ cup of

plain cornmeal, a little salt, and enough water to form a smooth-working dough. Work in a little cotton to give the dough strength and consistency. Drop the dough into boiling water for 20 minutes. Work into small balls when cool.

Carp Bait (2): Peel and grate two large potatoes, add ½ teaspoon salt, 1 tablespoon cornmeal, and enough wheat flour and cotton to produce a stiff batter. Roll balls one inch in diameter. Drop into boiling water until balls float to surface.

For Further Information

UNITED STATES

U.S. Department of the Interior
Fish and Wildlife Service
Bureau of Sport Fisheries and Wildlife
Division of Fish Hatcheries
Washington, D.C. 20240

CANADA

Department of Sport Fisheries,
Wildlife and Conservation
Ottawa, Quebec Province, Canada

19

Cooking Wild Game

Nothing of importance has been added to the killing, dressing or cooking of wild game in the last two hundred years. What has changed is the reasons we give for killing it. It is now classified as "sport."

The early mountain men and explorers who pioneered our wild frontier for those who later settled it lived almost exclusively on wild game and Indian maize (corn). As a result, they were a far hardier breed than the farmers who tamed the frontier and ate what they raised in garden and farmyard.

Animals in the wild are, for the most part, vegetarians, living off plants, seeds, fruits and berries. They did not then—nor do they now—live in crowded conditions and are suprisingly sanitary in habit.

People who eat game for the first time usually attempt to compare it with beef, pork, poultry or other domestic meats, but wild game has its own distinctive flavor which is not comparable with the meat of any other group of animals. With proper handling and preparation, any wild game recipe in this chapter will give you a distinctive, flavorsome and economical meal. Furthermore, the pelts of fur-bearing animals command a ready cash market to supplement the income of many farmers.

Today more *small* animals are taken by trappers for their fur than are shot by sportsmen, with the exception of wild birds and possibly also rabbits and woodchucks. Contrary to the opinions of an ignorant public, *trapping is not inhumane.* Most traps used by modern trappers kill instantly.

After Killing, Dress the Game Immediately

Much meat taken by hunters is wasted because of insufficient care after the game is shot.

The necessity for dressing game immediately after shooting depends, of course, on the weather. Hunters agree that game birds may be carried through the day without cleaning. If the temperature is warm and the birds are to be kept more than one day without ice or refrigeration they should be drawn (gutted) as soon as possible, but the feathers should be left on until the hunter reaches home. After the bird is drawn the body cavity should be wiped dry, using clean leaves or grass if a cloth is not available. *Do not use water*. It is advisable for hunters or trappers to clean out the abdominal cavity of mammals the same day they are killed. Extra caution should be taken in dressing cottontail rabbits, and dogs should not be allowed to eat the viscera of these animals because some diseases and parasites are spread in this way. Splintered bones, pieces of feathers or fur and shot should be carefully removed from shot wounds in order to present game to the cook in a suitable condition. Many hunters carry a clean paper or cloth sack in their hunting coat in which they put game.

Most hunters skin game birds instead of plucking feathers, as it is easier and means less work for the cook. However, many game cooks state that some of the taste of the meat is lost by removing the skin. Since fur animals are harvested for their pelts, the carcass is a by-product which can be used for food. If the carcass is to be eaten, trap lines should be run at least once each day and the animal pelted and dressed soon after removal from the trap.

The question sometimes arises as to the necessity for bleeding animals after shooting. Usually after an animal has been shot, sufficient bleeding occurs so that additional bleeding is unnecessary. However, if the animal is caught in a trap or killed in such a manner that little or no bleeding has occurred, it would be wise to facilitate bleeding.

This may not be feasible with muskrat or other fur animals trapped for their pelts, since it is undesirable to make cuts through the fur which might decrease its value. The meat from such animals should be put in a cold place and soaked in salt water (one tablespoon salt to one quart water) 8 to 10 hours before cooking. It is generally recognized by experienced cooks that freshly killed game should be allowed to age in a cool place at least 24 hours from the time it is killed till it is prepared for eating.

Game animals lead an active and vigorous life. Game frequently forage over large areas for food, and they must be alert, tense and ready at all times to escape from many enemies. Their muscles are likely to be tougher, drier and less palatable than those of farm animals. Hence game meats usually require more attention when cooking than those of domestic animals which are confined, fed fattening foods and protected from hazards. As with domestic meats, however, young animals are tender and require little cooking.

Not *all* edible wild animals have been included in the recipes and suggestions for preparation in this chapter. Only the more important game and furred species have been considered. Snipe, porcupine, coots, marsh hens, and even birds such as grackles are edible. Banquets featuring crows as the main course are common in many Midwestern communities.

Pheasant

Pheasant meat is similar to, but drier than, chicken. Most chicken recipes are suitable for preparing pheasant. Cooking in a covered roaster helps retain the moisture. Any moist heat method for preparing fowl is desirable for pheasants.

ROAST PHEASANT

1. 6-8 servings
2. Temperature 350° F.
3. Cooking time 2 hours

1 pheasant
1 quart boiling water
3 stalks celery
1 onion
1 teaspoon salt
1/8 teaspoon pepper
4 strips bacon
1 cup water

1. Clean pheasant. Put in pan and pour boiling water over bird and into cavity.
2. Put the celery and onion in bird. Do not sew up.
3. Rub bird with salt and pepper. Place in roasting pan and put bacon over breast.
4. Add one cup water and roast in a moderate oven (350° F.) uncovered for 2 hours or until tender.

BAKED PHEASANT

1. 6 servings
2. Temperature 375° F., then 325° F.
3. Cooking time 30 minutes, then 1 hour 30 minutes

1 pheasant
1 teaspoon salt
1/8 teaspoon pepper
½ cup flour
2 tablespoons butter
1 cup hot water

1. Dress, clean and cut pheasant into six pieces.
2. Sprinkle with salt and pepper. Dip in flour. Place in a greased roaster. Dot with butter and brown in moderate oven (375° F.) for 30 minutes.
3. Add one cup hot water, cover and bake in slow oven (325° F.) for 1½ hours or until tender.

CURRIED PHEASANT

1. 6 servings
2. Cooking time 1¾ hours

1 pheasant
½ cup flour
3 tablespoons fat
2 medium onions, minced
1½ tablespoons curry powder
2 tablespoons flour
3 cups broth
1 sour apple or stalk rhubarb
2 teaspoons salt

1. Clean and cut pheasant into eight or nine pieces.
2. Roll in flour and cook in hot fat until brown, removing each piece as it browns.
3. Cook onions in same fat in which meat was cooked. Add the curry powder with the flour. Cook slightly, add broth and stir until it boils.
4. Replace the meat, add the apple or rhubarb and salt.
5. Cover and simmer for 1½ hours or until tender.

Wild Duck

Wild duck meat is dark and dryer than domestic duck. To retain or add moisture, it may be roasted with strips of bacon on the breast. Cooking in a covered roaster also helps to reduce dryness. Duck is usually served rare. For those who prefer duck well done, additional cooking time should be given.

ROAST WILD DUCK

1. 2 servings
2. Temperature 325° F.
3. Cooking time about 45 minutes

1 duck (1¼ lbs.)
2 cups quartered apples
1 slice onion
2 teaspoons salt
¼ teaspoon pepper

1. Clean duck and wash thoroughly.
2. Fill the duck with peeled quartered apples. Sew up and tie in shape.
3. Rub with a slice of onion, then with salt and pepper.
4. Roast uncovered in a moderately slow oven (325° F.), allowing 20 to 30 minutes per pound.
5. If desired, duck can be basted every 10 minutes with one cup orange juice. Basting is not required, however, at this low temperature.

BARBECUED DUCK

1. 4 servings
2. Cooking time ½ hour

2 large duck breasts
4 teaspoons lemon juice
1 teaspoon Worcestershire sauce
1 teaspoon tomato catsup

1 tablespoon butter
1 teaspoon salt
½ teaspoon paprika

1. Cut breasts from two large ducks.
2. Broil under flame until brown or about 10 minutes.
3. Baste frequently with the following barbecue sauce: lemon juice, Worcestershire sauce, catsup and butter.
4. When meat begins to brown, sprinkle with salt and paprika; continue to broil for 20 minutes or until done.

SMOTHERED WILD DUCK

1. 3-4 servings
2. Cooking time 1½ hours

1 duck
1 teaspoon salt
¼ teaspoon pepper
½ cup flour
½ cup fat
1 cup milk

1. Cut cleaned duck into six or seven pieces.
2. Season with salt and pepper and roll in flour.
3. Fry duck slowly in hot fat until brown on both sides, about 30 minutes, turning only once.
4. Add the milk, cover tightly and simmer slowly for one hour or until tender. (It may be baked in slow oven, 325° F.)

Rabbits

Some care should be taken in handling cottontail rabbits owing to the possibility of tularemia (rabbit fever). While extremely few cases of this disease occur, nevertheless care should be exercised. Hunters should avoid "sick-looking or queer-acting" rabbits. Those that arise slowly in front of the hunter or dog should be viewed with suspicion. Most human cases of tularemia have been contracted through cuts or other injuries on the hands. Therefore it is wise to use rubber gloves when cleaning rabbits. Cooking rabbit until well done kills the germs and makes the meat edible.

The cottontail hunter occasionally observes curious warty or horny growths on the skin of rabbits he has bagged. These are most common on the legs and head, although they may be found on any part of the body. Such growths cause no damage to rabbits, and hunters need not feel concerned about eating rabbits afflicted with this condition.

BAKED STUFFED RABBIT WITH CARROTS

1. Temperature 400° F.

Stuffing:
3 or 4 average potatoes boiled and mashed
2 tablespoons butter
1 teaspoon salt
½ teaspoon pepper
1 teaspoon dried summer savory
1 cup finely chopped celery

Meat:
1 rabbit
2 large carrots, quartered
Bacon or pork fat
1 or 2 cups hot water

1. For dressing, mash potatoes to make a pint; season with butter, salt, pepper, savory and celery. Fill body of rabbit with this stuffing and sew it up.
2. Place rabbit on rack of baking pan with legs folded under body and skewered in this position.
3. Place quartered carrots beside it on the rack.
4. Lay bacon over the back to keep flesh from drying out. Fasten in place with toothpicks.
5. Put pan in hot oven and after first 10 minutes pour a cup or two of hot water over body; continue cooking until tender.
6. Shortly before rabbit is ready, remove bacon and let the rabbit brown.

RABBIT DELIGHT

1 young rabbit
1 tablespoon fat
1 cup broth
¼ cup lemon juice
¾ cup orange juice
2 green peppers, chopped
½ cup mushrooms, chopped
1 tablespoon parsley, chopped
Pinch of ginger
Salt and pepper

1. Joint the rabbit and brown pieces in fat.
2. Add broth and other ingredients.
3. Cover and cook slowly until tender.
4. Season to taste.

FRIED RABBIT

2 wild rabbits
Lemon juice
Salt, pepper and nutmeg
Egg
Bread crumbs
Parsley
Green peas
Toast

1. Dress and disjoint two rabbits. Wipe clean and parboil 10 minutes in water containing lemon juice. Drain.
2. Season with salt, pepper and very little nutmeg.
3. Dip in beaten egg, then in very dry bread crumbs. Fry in deep fat. Have the fat hot enough so a one-inch cube of bread is brown in 60 seconds.
4. Drain free of fat by holding each piece on a fork over the flame. It makes them crispy and leaves no fatty taste.
5. Place pieces on a hot dish, garnish with parsley and serve with green peas on toast.

RABBIT A LA MODE: HASSENPFEFFER

1 rabbit
Water
Vinegar
1 onion
½ teaspoon salt
6 peppercorns
1 bay leaf
Salt and pepper
Flour
3 tablespoons fat
Sweet or sour cream

1. Clean rabbit and cut into small pieces. Place in crock or jar.
2. Cover with vinegar and water in equal parts.
3. Add onion, salt, peppercorns and bay leaf.
4. Soak rabbit for two days, then remove meat, keeping the liquid.
5. Sprinkle with salt and pepper and roll in flour.
6. Brown in fat; pour in vinegar water to the depth of ¼ inch.
7. Cover tightly and simmer until done. Do not boil at any time.
8. Remove rabbit from pot, thicken drippings and add sweet or sour cream to gravy.

FRICASSEED RABBIT

1 rabbit
Bacon
Flour
Butter or fat
Salt and pepper
Milk
Onion juice

1. Quarter the rabbit. Strip with strings of bacon sewed through pieces of meat.
2. Roll in flour and brown in butter or other fat.
3. Season with salt and pepper, add milk very slowly, just enough to keep it from sticking and cook covered until tender.
4. Make gravy in pan by adding flour. Flavor with onion juice if desired.
5. Variation: add sliced onions to cover meat, one cup sour cream; cook covered until tender.

BROILED PARTRIDGE

1 partridge
Flour
2 tablespoons flour
½ cup cold water
Pepper and salt
Butter
Toast
Bacon

1. Open partridge on back. If partridge is not tender, place in a small baking pan with ½ inch hot water and cover. Put in hot oven for 15 minutes.
2. Roll in flour; lay on broiling irons, breast down, until tender.
3. Make gravy of 2 tablespoons flour in cold water, with pepper, salt and melted butter.
4. Stir in the liquid in which the birds were parboiled.
5. Serve with toast and bacon and with gravy, if preferred. Or slash birds in breast three times when done. Put a little butter, salt and pepper in each slash, place on toast, then pour liquid from pan over them.

Woodchuck

The muscles of woodchuck are dark and thick, but the meat is mild in flavor and does not require soaking. If the woodchuck is caught just before he begins his winter sleep, there is an insulating layer of fat under the skin. The excess fat should be removed, but it is not necessary to remove all the fat as its odor and flavor are not objectionable. However, it is advisable to parboil the meat of older animals before roasting or frying.

FRIED WOODCHUCK

1. 6 servings
2. Cooking time 1¼ hours

1 woodchuck
1 tablespoon salt
1 cup flour
3 tablespoons fat

1. Clean woodchuck and cut into six or seven pieces.
2. Parboil in salted water for one hour.
3. Remove from broth, roll in flour and fry in hot fat (deep fat may be used) until brown.

WOODCHUCK MEAT PATTIES
WITH TOMATO SAUCE

1. 8-9 patties
2. Temperature 325° F.
3. Cooking time 1¼ hours

1 woodchuck
1 cup bread crumbs
¼ cup ground onion
1 teaspoon salt
1/8 teaspoon pepper
2 eggs
3 tablespoons fat
1 cup catsup
¼ teaspoon Worcestershire sauce

1. Clean woodchuck. Remove meat from the bones and grind.
2. Add ½ cup crumbs, onion, salt, pepper, one beaten egg and one tablespoon melted fat. Mix thoroughly.
3. Shape into patties and dip into one beaten egg, then into ½ cup crumbs, and fry until brown in 2 tablespoons hot fat.
4. Add catsup and Worcestershire sauce and bake in a slow oven for one hour.

WOODCHUCK MEAT PIE

1. 6-8 servings
2. Temperature 400° F.
3. Cooking time 1½ hours

Meat:
1 woodchuck
¼ cup onion
¼ cup green pepper
½ tablespoon minced parsley
1 tablespoon salt
1/8 teaspoon pepper

4½ tablespoons flour
3 cups broth

Biscuits:
1 cup flour
2 teaspoons baking powder
¼ teaspoon salt
2 tablespoons fat
¼ cup milk

1. Clean woodchuck and cut into two or three pieces. Parboil for one hour.
2. Remove meat from the bones in large pieces.
3. Add onion, green pepper, parsley, salt, pepper and flour to the broth and stir until it thickens.
4. If the broth does not measure 3 cups, add water.
5. Add the meat to the broth mixture and stir thoroughly.
6. Pour into baking dish.
7. For the biscuits: sift the flour, baking powder and salt together. Cut in the fat and add the liquid. Stir until the dry ingredients are moist. Roll only enough to make it fit the dish.
8. Place dough on top of meat, put in a hot oven and bake 30 to 40 minutes or until dough is browned.

Opossum

Opossum meat is rather light, fine-grained and tender. Soaking is not necessary. Excess fat should be removed, but it is not necessary to remove all fat because it does not have an objectionable flavor or odor.

ROAST OPOSSUM

1. 6-8 servings
2. Temperature 350° F.
3. Cooking time 2½ hours

Meat:
1 opossum
1 tablespoon salt
1/8 teaspoon pepper
6-8 slices bacon
1 quart water

Stuffing:
1 tablespoon fat
1 large chopped onion
Opossum liver (optional)
1 cup bread crumbs
¼ teaspoon Worcestershire sauce
1 hard cooked egg
½ teaspoon salt
¼ cup water

1. Rub cleaned opossum with salt and pepper.
2. Put fat in skillet and brown onion in it. Add the opossum liver and cook until tender. Add bread crumbs, Worcestershire sauce, egg, salt and water. Mix thoroughly and stuff opossum.
3. Truss as you would a fowl.
4. Place in roasting pan. Lay bacon across back. Pour one quart of water into pan.
5. Roast in moderate oven uncovered until tender (about 2½ hours).
6. Baste every 15 minutes.

OPOSSUM WITH TOMATO SAUCE

1. 6-7 Servings
2. Cooking time 2 hours

1 opossum
1 tablespoon salt
¼ teaspoon pepper
1 sliced onion
¼ cup fat
2 cups tomato catsup
½ cup water
1 teaspoon Worcestershire sauce

1. Disjoint and cut an opossum into six or seven pieces. Place in a deep pan and cover with water.
2. Add the salt, pepper and onion to the cooking water and cook 1½ hours or until tender.

3. Melt fat in a thick skillet and brown on one side. Turn and immediately pour the catsup and water over meat. Add the Worcestershire sauce. Simmer 30 minutes.

OPOSSUM MEAT PATTIES WITH CATSUP

1. 8-9 medium patties
2. Temperature 325° F.
3. Cooking time 1¼ hours

1 opossum
1 tablespoon salt
¼ teaspoon pepper
¼ cup bread crumbs
¼ cup onion, chopped
1 egg
¼ cup milk
3 tablespoons fat
1½ cups tomato catsup

1. Clean the opossum. Cut meat from bones and run through meat grinder.
2. Add salt, pepper, crumbs, onion, beaten egg and milk. Mix thoroughly.
3. Shape into patties and fry in hot fat until brown.
4. When patties are browned on both sides, pour the catsup over them and place in a slow oven (325° F.) for one hour.

Raccoon
Raccoon meat is dark. The fat is strong in both flavor and odor, and most persons prefer to remove it before cooking. Raccoon is usually parboiled before roasting.

ROAST RACOON

1. 8 servings
2. Temperature 375° F.
3. Cooking time 3 hours

1 raccoon
2 tablespoons salt
½ teaspoon pepper
1 onion
3 carrots
1 cup broth

1. Clean raccoon and remove all fat. Parboil for one hour in water to which salt, pepper, onion and carrots have been added.
2. Place in a roasting pan, add one cup broth and roast uncovered in a moderately hot oven for 2 hours or until tender.

FRICASSEED RACCOON

1. 8 servings
2. Cooking time 2¼ hours

1 raccoon
2 tablespoons salt
½ teaspoon pepper
1 cup flour
¼ cup fat
2 cups broth

1. Clean raccoon and remove all fat. Cut into eight or ten pieces.
2. Rub with salt and pepper and roll in flour.
3. Cook in hot fat until brown, add the broth, cover and simmer for 2 hours or until tender.

RACCOON MEAT LOAF

1. 8 servings
2. Temperature 350° F.
3. Cooking time 1¾ hours

1 raccoon
½ cup cracker crumbs
½ cup ground onion
1½ tablespoons salt
½ teaspoon pepper
2 eggs
¼ teaspoon thyme
1 cup evaporated milk

1. Clean raccoon and remove the fat. Cut

meat off the bones and run through a food grinder.

2. Add the crumbs, onion, salt, pepper, beaten eggs, thyme and milk, and mix well.

3. Put into a meat loaf pan, set in a pan of hot water and bake in a moderate oven for 1¾ hours.

RACCOON GOULASH

1. 8 servings
2. Cooking time 3 hours

1 raccoon
3 tablespoons fat
3 cups broth
2 cloves garlic
2 bay leaves
1 teaspoon salt
¼ teaspoon cayenne pepper
3 tablespoons butter
3 tablespoons flour
2 tablespoons paprika
1 cup tomatoes

1. Clean raccoon and remove fat; cut meat into 1½-inch cubes.

2. Brown meat in hot fat; add the broth, garlic, bay leaves, salt and cayenne. Simmer 2½ hours.

3. Cream the butter, flour and paprika together, combine with a little liquid from the goulash and add to goulash.

4. Cook until it thickens.

5. Add the tomatoes and cook for 30 minutes.

Muskrat (Marsh Hare)

The muskrat is a leading fur bearer. Its home is in water along the shores of lakes, in marshes and in streams. The bulk of its food is water or shore plants. While the animal is taken for its pelt, the carcass is edible and should not be discarded.

The flesh of the muskrat is dark red, fine-grained and tender. The meat should be soaked overnight in a weak salt solution (one tablespoon salt to one quart water) to draw out the blood. If the "gamey" taste of these animals is objectionable, soaking in the salt solution or in a weak vinegar solution (one cup vinegar to one quart water) will reduce the intensity of the taste.

FRIED MUSKRAT

1. 4 servings
2. Cooking time 2 hours

1 muskrat
1 egg yolk
½ cup milk
1 teaspoon salt
½ cup flour
3 tablespoons fat

1. Soak muskrat overnight in salted water (one tablespoon salt to one quart water). Disjoint and cut muskrat into desired pieces.

2. Parboil for 20 minutes, drain and wipe with a damp cloth.

3. Make a smooth batter by beating the egg yolk and milk, then add the salt and flour.

4. Dip the meat in the batter and drop into hot fat and brown.

5. When brown, reduce the heat, cover and cook slowly for about 1½ hours.

SMOTHERED MUSKRAT AND ONIONS

1. 4 servings
2. Cooking time 1¼ hours

1 muskrat
1½ teaspoons salt
¼ teaspoon paprika
½ cup flour
3 tablespoons fat
3 large onions, sliced
1 cup sour cream

1. Soak muskrat overnight in salted water

(one tablespoon salt to one quart water). Drain, disjoint and cut up.

2. Season with one teaspoon salt, add paprika, roll in flour and fry in fat until browned.

3. Cover muskrat with onions, sprinkle onions with ½ teaspoon salt. Pour in the cream.

4. Cover skillet tightly and simmer for one hour.

MUSKRAT MEAT LOAF

1. 6-8 servings
2. Temperature 350° F.
3. Cooking time 1½ hours

1½ pounds ground muskrat meat
2 eggs, beaten
1/3 cup dry crumbs
1 cup evaporated milk
¼ onion, minced or grated
¼ teaspoon thyme
1 teaspoon salt
¼ teaspoon pepper
1 teaspoon Worcestershire sauce

1. Soak muskrat overnight in salted water (one tablespoon salt to one quart water). Remove meat from bones and grind.

2. Mix ground meat thoroughly with other ingredients.

3. Place in meat loaf dish.

4. Place dish in pan containing hot water.

5. Bake in moderate oven (350° F.) for 1¼ hours to 2 hours.

BAKED STUFFED MUSKRAT WITH CARROTS

1. 4 servings
2. Temperature 400° F.
3. Cooking time 1 hour

1 muskrat
3 medium potatoes
2 tablespoons butter
1½ teaspoons salt
¼ teaspoon pepper
1 teaspoon dried summer savory
1 cup finely chopped celery
2 large carrots
3 slices bacon

1. Soak muskrat overnight in salted water (1 tablespoon salt to 1 quart water).

2. Cook and mash potatoes with the butter, season with ½ teaspoon salt, 1/8 teaspoon pepper, savory and celery.

3. Fill the muskrat with this stuffing and sew it up. Rub muskrat with 1 teaspoon salt and 1/8 teaspoon pepper.

4. Place on a rack in a roasting pan with the legs tied under the body.

5. Place two large quartered carrots on the rack beside the muskrat.

6. Place bacon on the back. Bake in a hot oven. After 10 minutes, pour two cups of hot water over the body and continue cooking for 45 minutes. Remove bacon the last 10 minutes so as to brown the back.

Squirrel

Squirrel meat truly makes a tasty meal. The flesh is medium red in color, tender and has a pleasing flavor. The slight "gamey" taste present in most game meats is almost absent in that of the squirrel. No soaking is necessary, and only the oldest and toughest animals will require parboiling for tenderness.

FRICASSEED SQUIRREL

1. 4 servings
2. Cooking time 3½ hours

1 squirrel
½ teaspoon salt
1/8 teaspoon pepper
½ cup flour
3 slices bacon
1 tablespoon sliced onion
1½ teaspoons lemon juice
1/3 cup broth

1. Disjoint and cut squirrel into six or seven pieces.

2. Rub pieces with salt and pepper. Roll in flour.

3. Pan fry with chopped bacon for 30 minutes.

4. Add onion, lemon juice, broth and cover tightly. Cook slowly for 3 hours.

5. Variation: Add one tablespoon paprika, 1/8 teaspoon cayenne, one sliced sour apple, 2 cups broth instead of bacon, and 1½ teaspoons lemon juice.

BRUNSWICK STEW

1. 4–5 servings
2. Cooking time 3 hours 10 minutes

1 squirrel
2 quarts boiling water
1 cup corn
1 cup lima beans
2 potatoes
½ onion
1½ teaspoons salt
½ teaspoon pepper
1½ teaspoons sugar
¼ cup butter
2 cups tomatoes

1. Clean squirrel and cut into six or seven pieces.

2. To the water add the squirrel, corn, lima beans, potatoes, onion, salt and pepper. Cover and simmer for 2 hours. Add the tomatoes and sugar; simmer for one hour. Add butter and simmer for 10 minutes.

3. Bring to a boil and remove from fire. Add additional salt and pepper as desired.

ROAST SQUIRREL

1. 4 servings
2. Temperature 350° F.
3. Cooking time 1½ hours

1 squirrel
1½ teaspoons salt
¼ teaspoon pepper
1½ tablespoons lemon juice or tarragon vinegar
1 cup bread crumbs
¼ cup cream
1 cup button mushrooms
1 tablespoon onion juice
1 tablespoon melted fat
2 cups brown meat broth

1. Clean squirrel thoroughly. Rub with a mixture of one teaspoon salt and 1/8 teaspoon pepper, then with lemon juice or tarragon vinegar.

2. Soak bread crumbs in the cream to moisten them.

3. Add mushrooms (chopped), remainder of salt and pepper, and onion juice.

4. Stuff squirrel with this mixture, sew and truss as for a fowl.

5. Brush with melted fat and place in a dripping pan. Partly cover with the broth diluted with a cup of boiling water.

6. Roast 1½ hours in a moderate oven uncovered.

7. When the squirrel is well done, remove from pan. A gravy may be made from the liquid in the pan.

SQUIRREL PIE

1. 6-8 servings
2. Temperature 350° F.
3. Cooking time 1¾ hours

1 squirrel
3 tablespoons flour
½ tablespoon minced parsley
1 teaspoon salt
1/8 teaspoon pepper
½ cup fresh cut mushrooms
2 cups stock or milk

Biscuits:
2 cups flour
4 teaspoons baking powder
½ teaspoon salt
¼ cup fat
2/3 cup milk

1. Disjoint and cut squirrel into two or three pieces.

2. Cover with water and cook one hour.

3. Remove meat from bones in large pieces.

4. Add flour, parsley, salt, pepper and mushrooms to the stock. Cook until it thickens (5 to 10 minutes).

5. Add the meat and mix well. Pour into baking dish.

6. Make the biscuits by sifting the flour, baking powder and salt together. Cut in the fat and add the milk. Stir until all dry ingredients are moistened. Roll only enough to make it fit the baking dish.

7. Place dough on meat in baking dish.

8. Bake in moderate oven until dough is golden brown (30 to 40 minutes).

20

Venison

In frontier days venison was one of the main meat dishes of pioneering families. Youngsters from the backwoods took "veal" sandwiches to school in molasses-pail dinner buckets.

Today deer hunting is for rest and recreation. Venison, however, is as good as it ever was. If you are lucky enough to kill a deer, proper handling of the animal before it is made up into cuts of venison may determine how well you enjoy eating your hunting trophy.

Better Safe Than Sorry

Your rifle is a deadly weapon. Handle it carefully. Never point it at anything you do not expect to kill.

Sight in your rifle before going into the woods. That first shot may be all you will get, so make it a good one. The first shot, before the game is scared, is better than all the shots left in the magazine of your rifle.

Be sure your target is a deer before you shoot; make sure it is a legal deer. Nothing disgusts a good hunter more than finding illegal deer shot and left in the woods.

If you only wound your deer, follow its trail as far as possible and then go another hundred yards. Many deer are lost because the hunter did not follow the trail far enough. A mortally wounded deer will often travel a hundred yards or better and that is a long way, especially in a swamp.

Be sure he is dead. It is better to shoot him again in the neck than to watch him get up and bound off into the woods.

A shot halfway between the ear and throat should hit the jugular vein; this will help bleed the deer. If you plan to mount the head, however, pass up this shot.

If the deer is not head- or chest-shot or has not been dead for several minutes before you find him, bleed him well.

Swing the deer around so the head and shoulders are lower than the rump. Stand in back of the deer, close to the body, and reach over to do the sticking. Watch those feet; he may not be as dead as you think. Many a deer hunter has been hit by the flying front feet of a "dead" deer.

Quickly insert a knife 4-5 inches deep at the spot where the base of the neck joins the chest. This is along the side of the windpipe. Cut sidewise to sever the large arteries that are at the base of the neck.

The more blood that drains out, the better your venison will be when cooked.

If the deer does not bleed, don't worry about it. It is better to stick a deer and get no blood than not to try to bleed the deer properly.

Now tag the deer. Attach the tag to the antler. If it is a spike horn, poke a hole through the ear and attach the tag there.

Dressing Out: Roll the deer over on its back. Tie the legs to a tree, or block the shoulders with chunks of wood to keep the belly up. Or hold him in this position with your knees on the inside of his hind legs.

The first few cuts should be made carefully to avoid poking a hold in the intestines or paunch. Cut along the belly line from the pelvic bone to the chest. Be sure with this first cut that you cut only the hide.

Now cut through the belly muscles—watch out for the intestines and paunch. Once through the muscles, you can hold back the intestines with the back of your hand, while guiding the knife between the first two extended fingers—cutting edge up.

Some hunters, after cutting through the muscles, take the knife in their fist, with the point of the knife up, and shove the fist along the belly muscles. The fist is inside the body cavity and holds the intestines down while the knife cuts the muscles.

Now slide your sleeves up a little higher and reach up into the body cavity until you come to the diaphragm. That is the thin muscle floor that separates the chest cavity from the stomach cavity. Cut this muscle out, staying close to the ribs. Now reach in farther and cut the windpipe. The windpipe feels like a hose with wire rings in it to hold it open.

A steady pull with the left hand and a cut here and there with the knife will allow the lungs, heart, paunch and intestines to roll out of the body cavity.

Before you pull them out, cut the supporting cords around the anus until it is loose, and draw it into the body cavity.

Hang It Right: Next, hang the deer in a tree to

drain. Hang it head up and high enough so blood can drain out of the hole where you removed the anus.

A good way to hang your deer to drain is as follows:

Select a hardwood sapling (maple is one of the best) 2-3 inches in diameter. Bend the top down and tie it to the antlers. You may need to notch the sapling near the base so it will bend low enough for tying. Cut two forked sticks, about 8 feet long, and strong enough to support a portion of the weight of the deer. Place both sticks at the point where the deer is attached to the sapling. The sapling and the two forked sticks form a tripod. Raise the deer by moving the sticks closer together (first one stick, then the other) until the deer has cleared the ground. The bigger the bent sapling, the more it will help lift.

A hanging deer cools more rapidly than one that is in contact with the ground.

Now look inside and make sure you have removed the diaphragm, lungs, intestines and loose fat. Use a cloth to wipe the body cavity clean and dry.

When you get the deer to camp, hang it head up in a shady place where there is good air circulation.

Most of the strong, disagreeable flavor in venison is due to inadequate bleeding, delay or carelessness in dressing, or failure to cool promptly and thoroughly. Blood from a shot wound, spreading through the muscles, can also cause a disagreeable flavor.

The best venison has been skinned out immediately, cooled outside the hide and prepared in the camp while fresh.

Skinning Out: The sooner you remove the hide, the easier it will come off. All you need to skin a deer is a sharp knife. Hang the deer by the horns, low enough so you can reach the head. Put your knife under the hide where you finished opening the deer at the rib cage. Have the sharp blade out and the point pointing up. Slide the knife up the neck, under the skin all the way up to the head. With the knife under the hide, and the sharp part of the blade out, circle the neck just below the antlers and ears.

Now start removing the hide, working down and around the neck at the same time. If you have cut from the inside out, there will be very little hair on the meat. Pick off the few hairs found on the throat. Keep the hair side of the hide from touching the meat. Using your thumb, peel the hide as far as possible. Use the knife wherever cords or meat hold the hide in place. Case skin (skin without splitting hide) the front legs, pulling the hide down the leg without splitting the hide. Splitting the leg skin can be done after the hide has been completely removed.

You will have to use your knife around the knees. The leg can be severed at the knee with a knife, or you can saw off the leg just below the knee.

Keep rolling the skin down and case skin the hind legs. After the hide has been completely removed, cut off the bony part of the leg that you did not skin, and split the skin that covered the legs so it will lie flat on the ground, flesh side up.

Sprinkle the hide with salt. Use plenty of salt—it is cheap and preserves the hide. In a couple of days pour or wipe off the moisture accumulated on the hide. Resalt and roll into a bundle, hair side out.

Butchering the Carcass

Freezing is the best way to preserve venison. Cut up the carcass yourself; all you need is a saw, a cleaver and a sharp knife. Or let a professional butcher do the job for you.

There are two different ways to cut up a deer carcass. The first is to hang it up by the hind legs and saw it in half. It can also be chopped with a cleaver. Split the backbone the full length, including the neck. Then divide each half into cuts.

Another way to cut up a deer carcass is as follows: Cut along the line between rump-loin and round-flank, then split down the backbone and cut into parts as desired. Next, cut the line between loin-flank and rib-chop-breast and divide the flank and breast. The backbone is not split, and the loin and rib chops are cut, making a double chop of each piece.

The neck is removed without splitting the backbone. This can be cut up into neck roasts or

A white-tailed buck deer with his antlers in full velvet

boned out for deerburgers or stews. The arm and front shanks are then removed and the shoulder cut into roasts.

Using Venison Cuts

Hind and Fore Shanks: Bone out and cut into cubes for stew meat or grind for meat loaf or deerburgers.
Round: Cut into steaks and prepare according to one of the recipes that follow. If the meat is tough, make it into Swiss steaks.

A small tender leg can be roasted whole just like a leg of lamb. It can also be ground or made into stew meat. Or it can be cured and smoked.
Loin and Rib Chops: The loin is where you get the sirloin and porterhouse. Generally, they are called chops. The loin and rib chops make the best frying steaks. These cuts also make extra-choice roasts.
Shoulders: Here is where very good pot roasts come from. Cut to whatever size roast you want. The shoulder can also be corned or it can be boned out for stews or ground meat.
Rump: Bone for stews.
Neck: Cut up for pot roasts. Bone out for stews.
Flank and Breast: The flank and breast contain a lot of meat, but it is best used for soup, stews or ground meat.

Recipes for Venison

Venison is the most highly prized of game meats. As mentioned earlier, most of the objectionable or gamey flavor comes from careless handling of the deer once it has been killed.

Most of the gamey flavor is in the fat, so trim away all the fat possible. Venison fat, as it cools, also tends to be sticky or tallowy and to cling to the teeth and roof of the mouth.

Because venison is a dry meat, it is greatly improved by adding suet, butter, pork or bacon while cooking. When mixing venison with other meat for storage, using suet instead of pork will prolong the storage life of the venison. The suet does not become rancid as quickly as the pork fat.

BARBECUED DEERBURGERS

1. 8-10 servings
2. Temperature 400° F. to brown, 220° F. to finish
3. Cooking time 20 minutes to brown, ½ hour to finish

3 tablespoons fat or drippings
2 pounds ground venison
1 cup onion, chopped
1 cup celery, finely diced
½ large green pepper (chopped fine)
1 clove garlic, minced (optional)
½ cup chili sauce
½ cup catsup
1¾ cups water
2 teaspoons salt
¼ teaspoon pepper
2 tablespoons Worcestershire sauce
¼ cup vinegar
1 tablespoon brown sugar
2 teaspoons dry mustard
1 teaspoon paprika
2 teaspoons chili powder
2 tablespoons chopped parsley

1. Preheat frying pan. Add fat and melt. When hot, add meat, onions and celery.
2. Brown, stirring frequently. Spoon off excess fat.
3. Combine remaining ingredients, except parsley. Mix well and pour over meat.
4. Cover frying pan and simmer 30 minutes, stirring occasionally.
5. Add parsley. Serve between hot buns or over mashed potatoes, rice or noodles.

BARBECUED VENISON NO. 1

1. 5 servings

1 bottle (28 ounces) prepared barbecue sauce
1 cup catsup
2 tablespoons pickle relish
1 cup beef broth or pan juices from venison roast
1 small onion, chopped
2 stalks celery, chopped
2 pounds cooked rump roast of venison

1. Mix all ingredients except venison in large saucepan.

2. Cook over low heat for about 30 minutes or until sauce is thick.

3. Slice rump roast into sauce and simmer until meat is just heated through.

4. Serve on hard rolls crisped in oven.

BARBECUED VENISON NO. 2

1. Temperature 375° F.
2. Cooking time 20 minutes

2 onions, chopped
6 tablespoons salad oil
2 tablespoons sugar
2 teaspoons dry mustard
2 teaspoons paprika
1 cup water
½ cup vinegar
2 tablespoons Worcestershire sauce
2 drops Tabasco sauce (optional)
Sliced cooked venison

1. Brown onions in salad oil.
2. Add remaining ingredients except venison.
3. Arrange meat in casserole. Pour sauce over meat.
4. Bake for 20 minutes, or until sauce thickens.

VENISON AND CORN CASSEROLE

1. 8 servings
2. Temperature 375° F.

1 pound ground venison
4 celery stalks, diced
2 medium onions, chopped
1 can (10½ ounces) tomato soup
1 can (1 pound) cream-style corn
1 can (15½ ounces) kidney beans, drained
1 teaspoon garlic salt
Dash pepper
1½ tablespoons Worcestershire sauce
1 teaspoon chili sauce
1 package (3¾ ounces) corn chips

1. Preheat oven.
2. Brown ground venison in large heavy skillet. Add celery and onions. Cook and stir 3 minutes. Reduce heat.
3. Stir in soup, corn, beans and seasonings.
4. Pour into 2-quart casserole. Bake uncovered 20 minutes.
5. Top with corn chips. Bake 10 to 15 minutes more, or until chips are slightly toasted.
6. Variation: Omit corn chips. Bake 35 minutes. Serve with corn bread.

VENISON AND RICE CASSEROLE

1. Temperature 300° F.
2. Cooking time 1 hour

Lard or suet
2 pounds ground venison
Salt and pepper
2 cups celery, diced
2 cups onion, diced
1 green pepper, chopped
1 can mushroom soup
1 can chicken rice soup
1 cup uncooked rice

1. Melt lard or suet in large fry pan.
2. Add venison, salt, pepper, celery, onion and green pepper; cook until brown.
3. Combine remaining ingredients and pour over meat and vegetables. Bake for one hour.

TRUE VENISON JERKY

1. Cut lean strips of venison into pieces one to 1½ inches thick and about 5 inches long. Any cut can be used, but tender meat gives a better product. The loin, round and flank are often used.

2. Make a brine of ½ pound salt to one gallon water; store in granite canner, stone crock or plastic bucket. Add meat. Weight the meat so the liquid covers the surface, and let stand at least 12 hours.

3. Drain well and place on trays from

smoker. Transfer to smokehouse. Dry out and flavor with warm, not hot, smoke for five to fifteen days, depending on the size of the pieces. Use any nonresinous wood like maple, ash or apple. When completely dry, store meat in airtight containers. Jerky keeps indefinitely if all the fat has been removed before brining.

LIVER

1. Slice liver thinly.
2. Cover with water and bring to boil.
3. Pour off water.
4. Slice one large onion and sauté in 1¼ tablespoons of olive oil plus one tablespoon butter.
5. Fry liver with onion lightly. Season with salt and pepper. Squirt on a little lemon juice. Do not overcook or the liver will be tough.

MEAT BALLS

2 pounds ground venison
2 teaspoons salt
¼ teaspoon pepper
1 onion, chopped fine
1 cup celery, chopped
½ cup green pepper, chopped
4 eggs, slightly beaten
1 cup cracker crumbs, crushed
Tomato sauce or tomato juice
2 tablespoons shortening

1. Mix ground venison, salt and pepper, onion, celery, green pepper, eggs and cracker crumbs.
2. Shape into small balls and brown in shortening.
3. Pour tomato sauce or tomato juice over the meat balls.
4. Cover and allow to simmer for about one hour.

MEAT LOAF

1. Temperature 350° F.
2. Cooking time 1 hour

1 pound ground venison
½ pound ground pork
1 egg
½ cup dried bread crumbs
1 cup milk
½ tablespoon onion, chopped
1½ teaspoons salt

1. Beat egg; add bread crumbs and milk.
2. Mix thoroughly with meat.
3. Add onion and salt.
4. Put in greased pan and bake at 350° for one hour. Tomato and green pepper may be added for seasoning.

MEAT PIE

1. 6 servings

1 large onion, chopped
2 tablespoons shortening
1 pound ground venison
1 teaspoon salt
¼ cup canned green chilies, chopped (about 2 small peppers—may omit)
1 teaspoon oregano
1 can (8 ounces) tomato sauce
2 cups biscuit mix, prepared according to directions on package
½ cup American cheese, shredded

1. In 10-inch cast-iron frying pan, cook onion in shortening until wilted.
2. Add ground venison, salt, green chilies and oregano. Cook until brown, breaking meat with a fork.
3. Add undiluted tomato sauce and heat.
4. Pat out biscuit dough on a piece of floured waxed paper to a 10-inch circle. Cut into wedges and place, paper side up, on top of the "filling."
5. Peel off paper and bake in hot oven (425°) for 15-20 minutes until brown.
6. Turn upside down on a hot plate. Sprinkle with shredded cheese and slip under the broiler for a few minutes until cheese has melted.

MINCEMEAT

2 pounds cooked venison, chopped in food grinder
4 pounds apples, chopped
2 pounds raisins
4 cups brown or white sugar
3/4 pound chopped suet or butter
½ teaspoon cloves
1 teaspoon mace
½ teaspoon nutmeg
2 teaspoons salt
1½ teaspoons cinnamon
Cider

1. Add cider to cover mixture or use fruit juices or water with ½ cup vinegar.
2. Cook slowly until fruits are tender (about one hour).
3. Store in fruit jars or make into pies when cool.

CREOLE STEAK

1 large round steak
Flour
Salt and pepper
Fat
3 stalks celery, chopped
½ green pepper, chopped
3 large onions, chopped
1 cup tomatoes

1. Pound flour, salt and pepper into steak. Brown in fat.
2. Cover with celery, green pepper and onions. Add tomatoes.
3. Cover tightly and cook slowly until meat is tender (about 1¼ hours).

FRESH STEAK

Fresh steak can be good if properly prepared. This entire process takes about 3 minutes. The steaks will be rare, but tender and delicious.

1. Cut steaks one inch thick.
2. Put between waxed paper and flatten with a mallet or side of a cleaver until ¼ inch thick.
3. Have frying pan hot. Put in one tablespoon of butter.
4. Drop in the steaks, but keep shaking the pan so steak does not stick. Salt and pepper lightly while cooking.
5. Turn just once—all the time shaking the pan.
6. Serve hot with butter on top.

MARINATED STEAK

Steaks ¼ to ½ inch thick
1 quart vinegar
1 quart water
2 tablespoons salt
8 bay leaves and/or
8 whole cloves

1. Soak steaks 12-24 hours in liquid made from above ingredients.
2. Remove, drain and rinse in cold water. Dry on cloth or paper toweling.
3. Salt and pepper to taste.
4. Fry at medium heat in melted beef suet. (This is better than lard or oil.) Do not overcook.
5. Brush with butter and serve on a hot plate.

STEW

1. Temperature 350° F.
2. Cooking time 1½ hours

1½ pounds venison (any part cut in pieces for stew)
6 medium potatoes, cut in chunks
8 carrots, cut in chunks
3 celery stalks, cut in 2" lengths
1 package onion soup mix
1 small can tomato sauce
1 2-ounce can mushroom bits and pieces

1. Put venison, potatoes, carrots and celery

stalks in casserole and sprinkle with onion soup mix.

2. Add tomato sauce and mushrooms.

3. Cover tightly, either with lid or aluminum foil. Place in oven and bake 1½ hours.

STROGANOFF

½ cup oil (or other fat)
2 pounds venison cut in strips ½ inch by ½ inch by 2 inches
¼ cup flour
1 envelope onion soup mix
3 cups water
1 cup cream of mushroom soup
6 tablespoons catsup
Noodles, rice or mashed potatoes

1. While oil is heating, shake meat in flour to coat.

2. Sauté until browned.

3. Add onion soup mix and water. Simmer until venison is tender (one hour or more).

4. Add cream of mushroom soup and catsup.

5. Heat thoroughly and serve over noodles, rice or mashed potatoes.

Dry-Heat Methods (for Tender Cuts)

Roasting (round, loin, shoulder) 1. Season with salt and pepper. 2. Place on rack in uncovered pan, fat side up. 3. Do not add water—do not cover. 4. Extra fat may be added to venison. Bacon strips or beef suet may be laid across the top. 5. Roast in slow oven (300°-350° F.), allowing 20-25 minutes per pound. Turning the roast aids uniform cooking.

Broiling (steaks and chops) 1. Preheat the broiling oven. 2. Place steaks or chops on the broiling rack with top surface 3 inches below source of heat. 3. Broiler door should be open unless directions of range advise otherwise. Lower flame or heat if meat smokes or throws grease into flame. 4. Broil on one side until nicely browned and then turn to other side. For a one-inch steak, the time required will be 7 to 10 minutes for each side. 5. Season with salt and pepper, add butter and serve at once. (Try broiling in your fireplace over a bed of glowing coals.)

Panbroiling—Frying (steaks and chops) 1. Heat a heavy frying pan until it is sizzling hot. 2. A little butter added to the pan improves the flavor—otherwise rub the pan with a little suet or small amount of fat. Place the meat in the hot pan. 3. Brown both sides—turning only once. 4. For thick chops or steaks, reduce heat after browning to finish cooking clear through meat.

21

Wild Edible Forest Nuts, Fruits, Seeds, and Mushrooms

Daniel Boone and some of our other early pioneers went into the wilderness with only a rifle and a sack of salt, but lived in comfort on the game and wild food plants which the forests afforded. To follow Boone's example, you should know the nuts, fruits, seeds, herbs, roots and wild mushrooms that can be eaten to supplement your daily diet.

Pecans, Hickory Nuts, Walnuts and Butternuts

Foremost among the many kinds of native nut trees are a dozen species of hickory native to the eastern section of the United States, with nuts varying in size, thickness of shell and taste. The more important nuts are shagbark hickory, shellbark hickory, mockernut hickory and red hickory. There are also bitternut hickory and water hickory, inedible because of their bitter meat. Then there are black walnut and butternut, well known to residents of the East. Four other kinds of walnuts grow in the Southwest and in California. The Indians made nut oil for cooking by pounding the kernels of hickory nuts, walnuts, butternuts and others, then boiling in water and skimming off the oil. They also made bread from chestnuts, beechnuts, pecans and other nuts.

The triangular, thin-shelled, oily beechnuts of our eastern forests are small but delicious; after roasting they can be used as a substitute for coffee. Our native chestnut trees are not as numerous as they once were, having been virtually exterminated in this country by the chestnut blight during the last one hundred years.

Chinquapins, represented by a few kinds of small trees and shrubs in the Southeast, have similar though smaller nuts. Golden chinquapin, of the Pacific coast region in Oregon and California, also has small, hard-shelled sweet nuts. Two shrubby species of filberts, or hazelnuts, occur in the eastern forests and another in the Pacific states.

The large seeds, or pine nuts, of several species of pines in the western United States are delicious, though the smaller seeds of various other native pines can also be eaten. Pine nuts, which are borne in cones requiring two years to mature, long served the western Indians as food, raw or roasted. First among our truly wild nut trees is the piñon, a dwarf pine forming large areas of semidesert and woodlands chiefly in New Mexico, Arizona and southwestern Colorado. Each autumn the Navajo Indians and other residents harvest more than a million pounds of these nutritious oily nuts about one-half inch long.

Singleleaf piñon, a related species of the Great Basin and California, has slightly larger piñon nuts, or pine nuts, which have a mealy taste. The Digger pine of the California foothills, named for the Digger Indians, has edible seeds and very large cones. A related species in the same state, Coulter pine, also has large seeds. Limber pine, of the Rocky Mountain region, and sugar pine, of California and Oregon, have small edible seeds. In the Southeast the small seeds of longleaf pine are particularly good if roasted.

Though it is a common belief that acorns are fit only for feeding hogs, many species can be made edible and nourishing for people as well. The Indians gathered and stored quantities of acorns, which were ground into meal and baked into an unleavened pasty, nutritious bread.

The tannin, which causes the bitter and astringent taste in raw acorns, was removed by soaking in water and filtering, or by boiling and leaching with ashes. Acorns were also eaten roasted. As a rule, acorns of the species in the white oak group are less bitter, and better for food, than those in the black oak group.

The following kinds of white oaks in the eastern United States have sweetish acorns that can be eaten roasted or raw or made into bread: white oak, swamp white oak, chestnut oak, swamp chestnut oak, chinquapin oak, dwarf chinquapin oak, live oak, post oak and burr oak.

Indians and Mexicans in the Southwest eat the sweetish acorns of emory oak, which are known by the Spanish name of bellotas—pronounced bay-YOH-tahs. In California the Indians hoarded acorns of California white oak, canyon live oak, California live oak and California black oak. Acorn meal is prepared by grinding the shelled kernels in a food chopper. The bitter tannin is removed by spreading the meal about one-half inch thick on a porous cloth and then pouring on hot water to percolate through, repeating once or twice as needed. Or the kernels may be boiled for 2 hours before grinding and then soaked in hot water with occasional changes until the bitter flavor is lost. After being dried and parched in an oven, acorn meal is used like corn meal in recipes for bread or muffins, alone or mixed with equal parts of corn meal or wheat flour.

One of the best wild fruit trees is the common persimmon, which is also native to the eastern United States, except along its northern border. As everyone knows, persimmons are very puckery when green and do not become edible until thoroughly ripe and soft, usually late in the fall. However, frost is not required. If a persimmon is eaten before maturity, the tannic acid in it has a strongly constrictive effect. Cooked persimmons can be prepared as a pudding, while another edible species, Texas persimmon, is native to southern Texas.

Pawpaw fruit is a northern representative of the custard-apple family of the tropics, which becomes fully ripe in late autumn and is 3 to 5 inches long and dark brown, resembling a short banana.

The rose family, to which many of our cultivated fruits belong, is well represented among the wild fruit trees. Several kinds of wild crab apples in the eastern United States have miniature apple fruits which make delicious jellies. Another species is found on the Pacific coast. Hawthorns, with numerous species in the eastern United States and a few in the West, are spiny shrubs or small trees with reddish fruits resembling tiny apples. Some are edible and can be made into preserves and jellies. Several species of serviceberries, small trees or shrubs also known as "juneberries," occur across the country. The sweet purplish fruits one-quarter to one-half inch in diameter are excellent raw or in pies. Various shrubby and tree species of wild plums, commonest in the East, have delicious fruits which are eaten raw or made into preserves, jellies and pies. Likewise, in different parts of the country, the fruits of several kinds of wild cherries and chokeberries are used for jelly and jam, though they usually have a sour or puckering taste when raw.

Red mulberry, a tree of the eastern half of the

country, has sweet, juicy, purplish fruits which are eaten fresh or made into pies, jellies and jams. The small pulpy "berries" of hackberries and sugarberry, with a few species in various regions, are sweet and suitable for nibbling. The purplish or bluish black fruits of "elderberries" are often made into pies, jams and jellies in different parts of the country, though the bright red berries of other kinds are bitter.

Southern Florida has native tropical fruit trees, such as darling plum and coco plum. Jelly is made from the fruits of sea-grape trees. Coconuts, which are widely distributed on the shores throughout the tropics, are naturalized in southern Florida. Native palms with small, edible fruits include cabbage palmetto in the southeastern states, Texas palmetto in southern Texas, and California palm in California and Arizona.

Kentucky coffee tree derives its name from the use of the seeds ground and roasted as a coffee substitute by pioneers about the time of the Revolutionary War. These large seeds were eaten roasted by the Indians. Some kinds of sumac, which occur across the country as shrubs or small trees, have red fruits pleasant for nibbling or sucking because of the very sour, sticky hairs. "Indian lemonade" is made by bruising and soaking (but not boiling) these fruits in water.

Then there is sassafras tea, prepared by brewing small pieces of the roots and root bark from sassafras trees, a common beverage and spring tonic especially in the Southeast. Sweet birch in the Northeast has aromatic oil of wintergreen in the bark, twigs, leaves and buds, from which a tea can be prepared. Yellow birch is similar but less aromatic. The leaves of a few species of evergreen hollies in the southeastern United States, including yaupon, dahoon and inkberry, contain small quantities of caffeine and are used in preparing a mildly stimulating drink.

Dried, powdered young leaves, tender stems and buds of sassafras are used to thicken and flavor soups. The dried, aromatic leaves of swamp bay are a favorite for flavoring soup and other dishes in the Southeast.

Buds of basswood are edible raw or cooked. The flower buds, flowers and young pods of redbud have been fried in butter, while the flowers can be used for salads and pickles. The large flower buds of yuccas can be prepared by roasting.

The inner bark of pines, spruces and firs was eaten by the Indians, sometimes ground and prepared into bread, and it can be used as an emergency food. Some deciduous trees reported to have inner bark that can be eaten if necessary are aspens, cottonwoods, alders, birches, hickories, slippery elm, American elm, sassafras and basswoods.

Wild Mushrooms

"Elm" mushrooms can be found in the form of clumps on dead and dying trees, in temperate regions of the United States, including Alaska, and in Canada.

The "oyster" mushrooms, one species of which has pink spores and the other white, have white, gray or light brown caps up to 9 inches in diameter. Like the "elm" mushroom, they grow in clumps.

The "elm" mushroom differs from the "oyster" mushroom mostly because of its long and curved stem. The younger mushrooms of this species may be French fried, made into soup or sautéed in salad oil. They may also be frozen should you be diligent enough to harvest a surplus.

The "Inkies," "Shaggy Mane" and "Early Inky" are probably the most common species found in the temperate zones of this country and Canada. Easily recognized by their black spores, which turn to a black inky mess when they become old, poisonous and inedible, they are absolutely delicious when sautéed in sweet butter and their own natural juices.

Pictured on the next two pages are some of the most delicious mushrooms to be found.

What to Do with Mushrooms

Now that you have learned to identify edible mushrooms, here are a few hints to help make the best of your find.

Store mushrooms in the refrigerator. Do not clean them if they are to be used within a day.

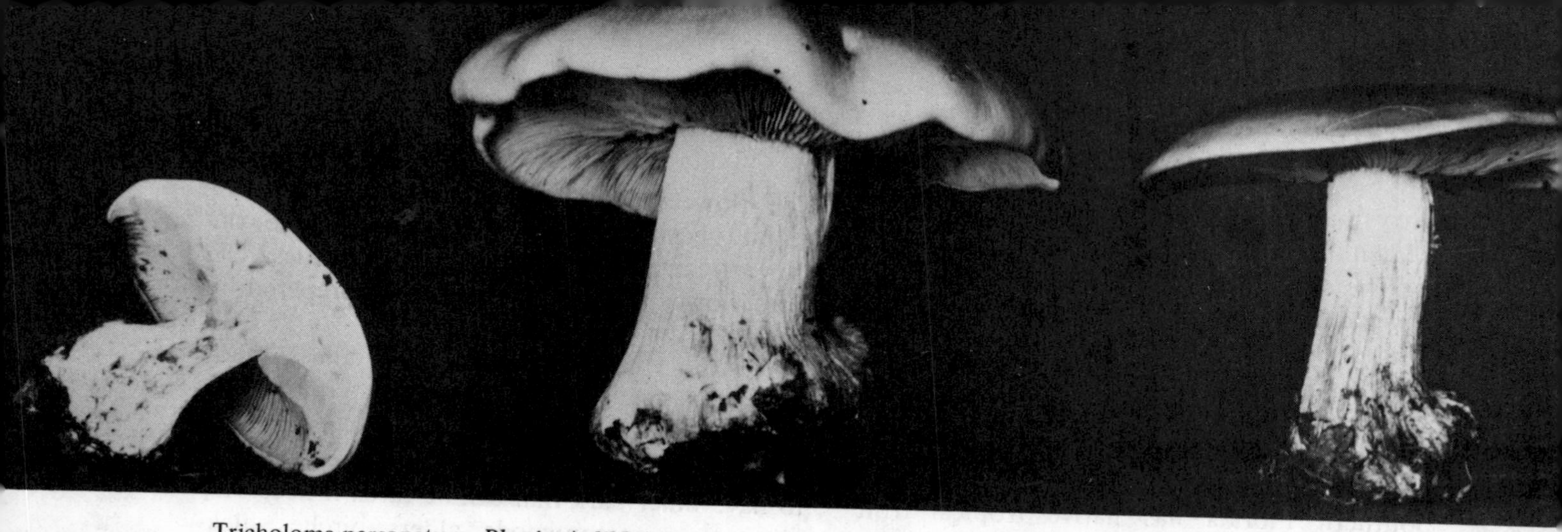

Tricholoma personatum, *Blewits (edible). In this species the cap is convex, expanded, slightly depressed, fleshy moist, pale tan, tinged gray or violet; young plants may be entirely violet, margin downy, involute; the flesh is whitish; the gills are crowded, rather broad, rounded behind, nearly free, violaceous, changing to dull reddish-brown; the stem is stout, sub-bulbous, fibrillose, solid, colored like cap or lighter. The cap is 2 to 5 inches broad; the stem is 1-1/2 to 2-1/2 inches long and 1/2- to 3/4-inch thick.* Tricholoma personatum *is found quite commonly in the late summer and fall months growing on the ground in woods and open places. This is one of the most acceptable edible species. They are often confusing to the amateur, but may be distinguished from each other by the fact that in* T. nudum, *the margin of the cap is naked and is thinner than in* T. personatum. *Also* T. nudum *is more slender than* T. personatum *and has deeper coloration on the cap and gills.*

Marasmius oreades, *Fairy-Ring Fungus (edible). In the fairy-ring mushroom the cap is convex, then plane and slightly umbonate, tough, smooth, brownish buff, later cream-colored, margin; when moist may be striate; the gills are broad, free, distant, unequal, creamy white; the stem is tough, solid, equal, villose in the upper part, smooth at the base. The cap is 1 to 2 inches broad; the stem is 2 to 3 inches long and 1-1/2 lines thick. Many allusions in literature undoubtedly refer to this interesting mushroom, and many fairy stories have happy associations with it. Its frequent occurrences on grassy places—lawns, pastures and golf courses—insure its wide acquaintance. It is to be found from early spring until autumn. This is a popular edible species and if once learned should always be recognized. It may be preserved for winter use by drying and is also well adapted for pickling.*

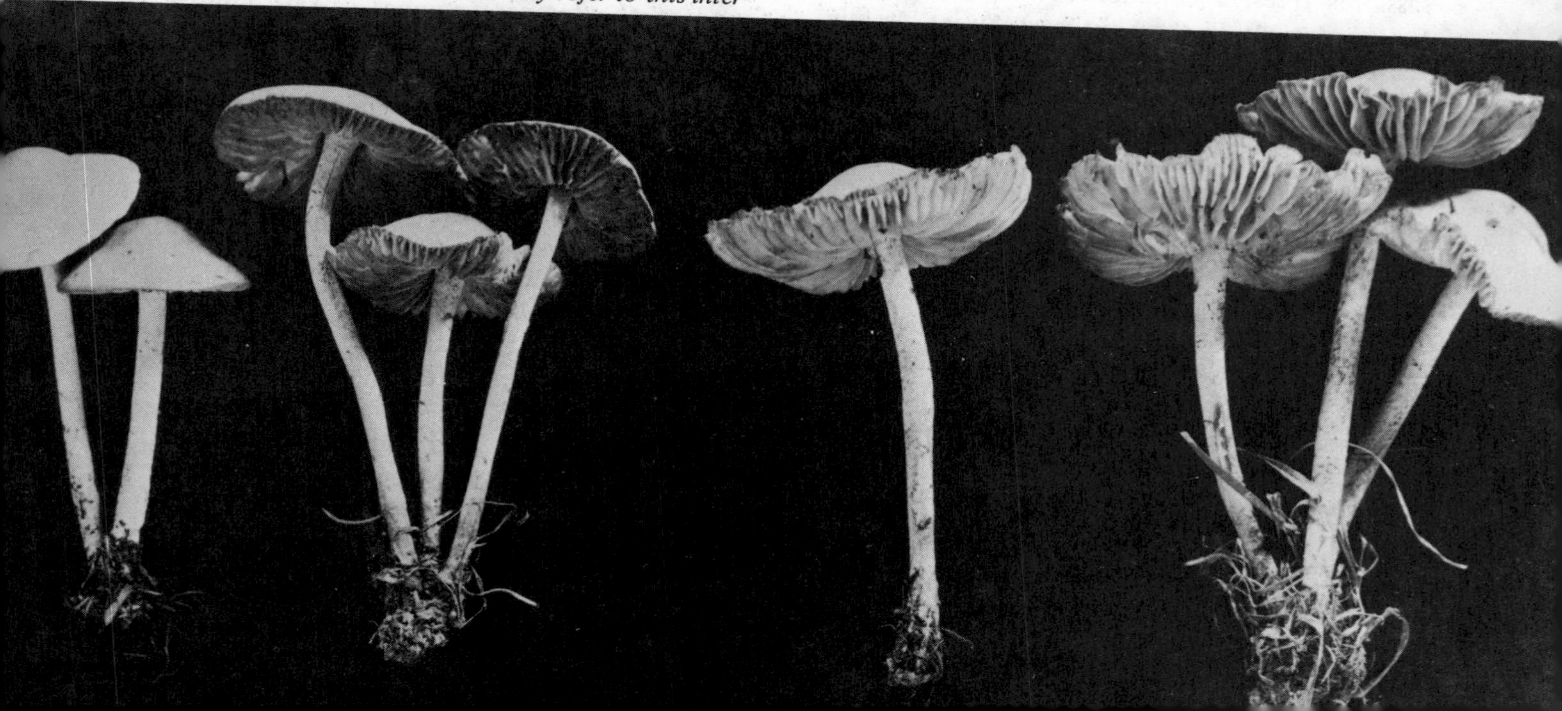

Cortinarius lilacinus *(edible). In this species the cap is firm, hemispherical, then convex, minutely silky, lilac-colored; the gills are close, violaceous changing to cinnamon; the stem is solid, stout, distinctly bulbous, silky fibrillose, whitish with a lilac tinge. The cap is 2 to 3 inches broad; the stem is 2 to 4 inches long. This is a comparatively rare but very beautiful mushroom and an excellent edible species. It is to be found in mossy or swampy places.*

Tricholoma equestre, *Equestrian Tricholoma (edible). In this species the cap is convex, becoming expanded, margin incurved at first, then slightly wavy, viscid, pale yellowish with a greenish or brownish tinge; the flesh is white or slightly yellow; the gills are sulphur yellow, crowded, rounded behind, and almost free; the stem is stout, solid, pale yellow, or white. The cap is 2 to 3 inches broad; the stem is 1 to 2 inches long and 1/2- to 3/4-inch thick. This species has a fairly wide geographic distribution and occurs very abundantly in Virginia, Maryland and the District of Columbia from the middle of November until about Christmas. It is to be found in pine woods, where it forms irregular or incomplete fairy rings. The plants exert considerable force in pushing their way out of the ground through the dense mat of needles, which often adheres so closely to the caps that slight elevations are the only indications of the presence of the mushrooms.* Tricholoma equestre *is a very excellent edible species and is delicious when fried or made into soup. The latter resembles turkey soup, but possesses a more delicate flavor.*

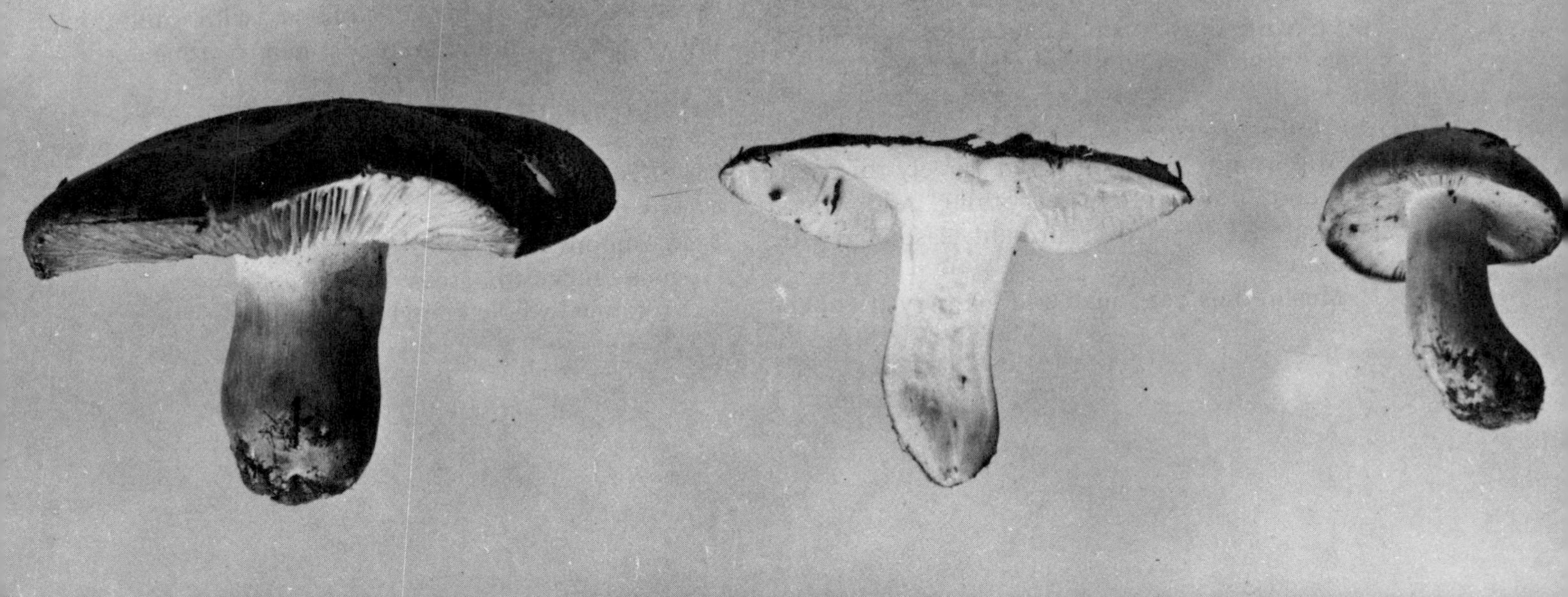

For storing more than one day, clean and cook first.

To clean, cut off the soiled roots and peel the rough caps. Rub off the fuzz or scales with a damp cloth. The mushrooms shown are edible. Shun other species. They may be poisonous.

Preservation

Freezing: Select large firm mushrooms; clean and sort according to size. Cut into pieces not larger than one inch across. To freeze, follow the directions for freezing vegetables that came with your home freezer.

Steam: To prevent darkening, soak the mushrooms 5 minutes in a solution of one pint of water with either one teaspoon of lemon juice or ½ teaspoon of citric acid powder. Drain and steam in a covered steamer: 5 minutes for large mushrooms, 3½ minutes for smaller ones. Dip in cold water to cool. Drain, pack, seal and freeze at once, leaving ½ inch head space.

Pan Fry: Heat mushrooms in a small amount of table fat until almost done.

Canning: Shaggy Manes, "Inkies" and meadow mushrooms *can* best. Drop cleaned mushrooms in boiling water containing one tablespoon of vinegar and one teaspoon of salt per quart. Simmer 3 to 5 minutes. Fill jars to ½ inch from the top with mushrooms and cover with fresh boiling water. Add ½ tablespoon lemon juice to preserve the color. Seal according to type of lid used. Or process in pressure cooker for 25 minutes at 10 pounds of pressure.

Drying: Cut large mushrooms in ½-inch slices; discard the tough stems. Leave small mushrooms whole. Spread out in layers not more than ½ inch deep. Dry to leathery dryness in a 150° oven with door open a crack. Place in plastic bags or foil. Store in a cool, dry place.

Mushroom Recipes

Mushrooms are composed largely of water and cook down to almost nothing. Add little or no water. Do not oversalt; add seasonings after cooking.

Mushrooms get tough and leathery if cooked too long; use a simmering temperature and short cooking time.

STEAMED OR COOKED MUSHROOMS

Add very little water. Cover tightly and simmer over low heat 10 to 15 minutes. The cooked mushrooms are ready to add to other food or to store in refrigerator or other very *cool* place within a few days.

SAUTÉED MUSHROOMS

Cook in a small amount of butter 3 to 8 minutes. Season with salt and pepper. Sautéed mushrooms are then ready to serve on toast or to add to sauces, gravies, cooked vegetables or meat.

Sautéed mushrooms will keep several days in the refrigerator in a covered dish.

BROILED MUSHROOMS

Brush caps with melted butter and place in a shallow pan about 3 inches from the heat. Season with salt, pepper and nutmeg. Broil 8 to 10 minutes. Serve on a variety of dishes.

RAW MUSHROOMS

Cut in slices and serve with well-seasoned salad combinations.

CREAMED MUSHROOMS

Add a cup of sautéed or cooked mushrooms to white sauce to serve alone or with cauliflower, peas, fish and seafoods, ham or tongue.

MUSHROOM GOULASH

1 large onion
1 quart mushrooms
2 tablespoons fat
1½ cups canned tomatoes
½ cup canned whole kernel corn

1 bay leaf
1 tablespoon chopped parsley
Salt and pepper to taste
½ cup sour cream

1. Dice and brown a large onion in fat.
2. Add mushrooms and simmer 5 minutes.
3. Combine remaining ingredients except sour cream and add to pan.
4. Add sour cream and reheat just before serving.

MUSHROOM MEAT SAUCE

2 cups mushrooms
Butter
3 tablespoons flour
Meat drippings

1. Sauté mushrooms in butter.
2. Add 3 tablespoons flour. Stir well.
3. Add enough water to meat drippings to make 2 cups. Add this to flour mixture.
4. Stir and cook until smooth.
5. Pour mushroom sauce over ham, turkey, fish, steaks or roast when ready to serve.

PICKLED MUSHROOMS

2 cups small button mushrooms
1 teaspoon salt
2 bay leaves
1 clove garlic
1 cup vinegar
1 sprig tarragon

1. Add salt, bay leaves and garlic to mushrooms.
2. Cover with hot vinegar boiled with tarragon.
3. Let stand three days before using.

Any of the above recipes may also be used with domestic (cultivated) mushrooms, fresh or canned.

22

A Treasury of Sourdough

A good many native and adopted Alaskans are ready to give battle to anyone who argues that sourdough did not originate in that state. I can swear to the truth of this because, although never coming to blows, I have engaged in many battles over the subject. A number of other Alaskans, some of whom I know to be authorities on the subject, claim only that the term *sourdough* was coined to separate the native Alaskan from the Cheechakos (newcomers) who swarmed into what was then still a territory after the Klondike (River) goldstrike (Yukon Territory, Canada) in the summer of 1896.

These newcomers from all over the world swarmed through Alaska in tens of thousands en route to the "strike." Gold was discovered about three years later on the beaches of Nome, and Alaska itself became the new "El Dorado," to which tens of thousands of "get-rich-quick" seekers were immediately attracted.

To miner, trapper and homesteader, native and newcomer to Alaska, sourdough was the "staff of life." It is possible they could not have survived without it.

As to its origins, I can only say that it was known to Americans for two centuries or more before Alaska was purchased from Russia by the United States; it was brought to these shores by early settlers from Europe.

In any case, sourdough is a yeasty starter for leavening bread, pancakes, waffles, muffins and cake. Unlike yeast, it will last forever. There are families in Alaska and elsewhere who can trace the ancestry of their sourdough starter back a hundred years or more. As for performance, it makes a dough so light that certain liars I know claim they must shut the window while making bread to prevent the dough from floating away.

Being an enthusiastic but truthful sourdough fanatic, I can solemnly swear that sourdough is the lightest there is, but if any dough I ever made floated in the air, it must have landed again before I returned to the kitchen. At any rate, it's great stuff to have around the house for baking, and the recipes that follow are some of my own favorites.

Sourdough Starter

Use a glass or pottery bowl. Never mix ingredients with a metal spoon or store the finished sourdough in metal bowls or containers.

Mix well 2 cups of warm potato water (water

in which potatoes have been boiled) and one package of dry yeast or a fresh yeast cake. Place in a warm place overnight. In the morning remove ½ cup of the mixture to a pint glass jar that has been scalded in very hot water. Store the contents in a cool place. This ½ cup will be your original starter. The remaining mixture is now ready for immediate use. Fermentation has now been achieved and mix can be used in any of the following recipes that call for 2 cups of starter. If not used within day, mix should be discarded.

Things to Know About Sourdough Starter

A good starter gives off the odor of clean sour milk. After standing for several days the liquid will separate from the batter. This action is perfectly normal and the starter itself will stay fresh forever, if it is replenished after three or four days with more flour and water. This is true for the starter, or sponge, as some purists call it, that is to be used daily, or at least several times a week.

If you do not intend to use the starter frequently, it can be stored *deactivated* in several ways. (1) Remove it from the jar and put it in your home freezer. (2) If the outside temperature is below freezing, remove it from the jar to any type of bag. (3) Add enough flour to the mixture so that the dough can be easily rolled into a ball and place the ball into a sack of flour. The sponge will then go into a dry form; that is to say, the fermenting agent will enter the spore state, which will then keep the mix inert forever.

When you wish to bring the starter back to life, add enough lukewarm water to make it soft. If it has been frozen, place it in a warm place until fully thawed out and returned to room temperature.

SOURDOUGH BREAD

1. Temperature 375° F.
2. Cooking time 50 to 60 minutes

2 cups starter
4 cups white flour (sifted)
2 teaspoons sugar
1 teaspoon salt
2 tablespoons fat
½ teaspoon baking soda

1. Sift dry ingredients into bowl (*except baking soda*), making well in center.
2. Add fat to the starter and mix thoroughly.
3. Pour into well in flour. If necessary, add enough additional flour to make soft dough that will knead easily.
4. Knead on floured board for 15 minutes (or knead one hundred times).
5. Place in greased bowl and turn to coat dough.
6. Cover with towel and let rise in warm place until doubled in size (2 to 4 hours).
7. Dissolve baking soda in tablespoon of warm water and add to dough. Knead it in well.
8. Shape dough into loaves to fit breadpans and set aside to rise. When doubled, bake in oven for 50 to 60 minutes.

Note: You have used up your two cups of starter. Now you will be working with the ½ cup of starter you have in reserve.

SOURDOUGH WHEAT BREAD

2 cups starter (set previous night)
1 cup whole wheat or graham flour
1½ teaspoons salt
2 tablespoons sugar
1 cup white flour

1. Combine ingredients and mix well with wooden fork (dough will be sticky). Set in warm place for 2 hours or more.
2. Turn out on warm, well-floured board. Knead one or more cups of white flour into dough for 5 to 10 minutes.
3. Shape into round loaf and place in well-greased pie plate.
4. Grease sides and top of loaf and cover with towel. Let rise one hour or until doubled in size.
5. Bake in preheated oven at 450° F. for 10 minutes, then reduce heat to 375° F. and bake 30 to 40 minutes longer. Makes one large loaf.

6. If starter is very sour, add ¼ teaspoon baking soda to the dough while kneading it on the board.

CASSEROLE BREAD

Use the same ingredients and methods as for sourdough bread. Do not knead the dough, but beat it 2 minutes at medium-speed mixer setting or three hundred strokes by hand. Let stand in mixing bowl until double in bulk. Add ¼ teaspoon of baking soda. Mix ½ minute, turn into greased casserole or loaf pan and let stand for 40 minutes. Bake as for sourdough bread. Bread is done when crust sounds hollow when tapped with finger.

SOURDOUGH FRENCH BREAD

Use same ingredients and methods as in sourdough bread, but add one package of dry yeast or cake of fresh yeast to starter. Shape into two loaves by dividing dough in half. Roll each half into a 15- by 12-inch rectangle. Wind up tightly toward you, beginning with the wide side. Seal edges by pinching together. Place rolls diagonally on greased baking sheets which have been lightly sprinkled with cornmeal. Let rise about one hour or until doubled in bulk. Brush with cold water. Cut with knife to make one or two lengthwise or several diagonal ¼-inch deep slits across tops of loaves. Place in hot 400° F. oven with pan of boiling water. Bake 15 minutes. Remove from oven and brush again with cold water. Reduce oven temperature to 350° F. and bake 35 to 40 minutes until golden brown. Brush third time with cold water and bake 2 to 3 minutes longer. Makes two French loaves.

SOURDOUGH HOT CAKES

The night before baking, remove your ½ cup reserve starter and place in medium-size mixing bowl. Add 2 cups of water (warm) and 2 cups of flour. Beat well and set in warm place overnight. In the morning you will find starter has gained one half again in bulk. It will also be covered with bubbles. Set aside ½ cup in *cleaned* pint jar as reserve and store in cool place. Remaining starter will make hot cakes for three. To this starter, add:

1 or 2 eggs
1 teaspoon soda
1 teaspoon salt
1 tablespoon sugar

Beat with wooden fork, blending in all ingredients. Add 2 tablespoons melted fat. Bake on hot griddle, turning only once. Serve with maple syrup, molasses or honey and butter.

SOURDOUGH CHOCOLATE CAKE

½ cup starter (thick)
1 cup water
1¾ cups flour
¼ cup nonfat dry milk

Mix and let ferment 2 to 3 hours in warm place until bubbly and there is odor of clean sour milk. Add:

1 cup sugar
1 cup shortening
½ teaspoon salt
1 teaspoon vanilla
1 teaspoon cinnamon
1½ teaspoons baking soda
2 eggs
3 squares melted chocolate

Cream fat, sugar, flavorings, salt and baking soda. Add eggs one at a time, beating well after each addition. Combine creamed mixture and melted chocolate with sourdough starter. Stir three hundred strokes until blended. Pour into two-layer pans or one larger pan. Bake at 350° F. for 25 to 30 minutes. Cool and frost with butterscotch-chocolate frosting (follows) or other icing.

BUTTERSCOTCH-CHOCOLATE FROSTING

3 1-ounce squares unsweetened chocolate
¼ cup butter or margarine
½ cup light cream
2/3 cup brown sugar
¼ teaspoon salt
Vanilla
Confectioner's sugar

1. In saucepan combine chocolate, butter, cream, brown sugar and salt.
2. Bring to boil, stirring constantly; cook until chocolate is melted. Remove from heat.
3. Add vanilla and enough confectioner's sugar for good spreading consistency (about 3 cups).
4. Spread over sides and top of cake.

SOURDOUGH WAFFLES

Set the starter as for hot cakes. Make it slightly thicker. Let stand overnight. Remove the usual ½ cup of starter for the next time, and to the remaining starter add:

1 teaspoon salt
2 tablespoons sugar
1 teaspoon baking soda
2 eggs
¼ cup melted fat

Mix well and add fat just before baking. Bake according to directions that come with waffle iron.

23

The Last Word

No matter what you might have heard, rural folk are extremely hospitable and kind to newcomers, whether crew-cut or long-haired. They are endlessly helpful, though not usually inclined to accept you as "one of us" until you've proven yourself to their satisfaction; go out of your way to cultivate and cooperate with them in every way possible.

Listen to their advice, but don't always accept it at face value. Check their suggestions against the expert advice of your local county farm agent.

Farm Communes

Heathcote Centre School of Living, Rt. 1, Box 129, Freeland, Maryland, issues a listing of farm communes (Intentional Communities) at a price of 50 cents. Another good bet for information on communes is Twin Oaks, Louisa, Virginia 23093.

Farm Agencies

The two major farm agencies that have not been listed elsewhere in this book are:

Strout Realty
P. O. Box 2557
Springfield, Missouri 65803

United Farm Agency
612 West 47th Street
Kansas City, Missouri 64112

Publications

There are far too many regional farm publications to list here, but the small farmer should probably subscribe to the following:

Organic Gardening and Farming
Rodale Press
Emmaus, Pennsylvania 18049

Mother Earth News
P. O. Box 38
Madison, Ohio 44057

Green Mountain Post
P. O. Box 479
Brattleboro, Vermont 05301

Of course, *The Last Whole Earth Catalog* is highly recommended and can be purchased at any book store.

Mail Order Houses

Sears Roebuck & Company
303 East Ohio Street
Chicago, Illinois 60611

Montgomery & Ward
618 West Chicago
Chicago, Illinois 60610

Aladdin Industries, Inc.
703 Murfreesboro Road
Nashville, Tennessee 37210
(For hard-to-find wood-burning stoves and kerosene-burning lamps)

National Farm Publications

Grit
208 West Third Street
Williamsport, Pennsylvania 17701

Southern Unit Publications
P. O. Box 6429
Nashville, Tennessee 37212

These are recommended because they carry advertising for hard-to-get products and because they are published for the "small farmer."